BENGAL DISTRICT GAZETTEERS

24 - PARGANAS

BENGAL DISTRICT GAZETTEERS

Burdwan/ **J.C.K. Paterson**
Darjeeling/ **L.S.S. O'Malley**
Howrah/ **L.S.S. O'Malley**
Hooghly/ **L.S.S. O'Malley**
Santal Parganas/ **L.S.S. O'Malley**
Sambalpur/ **L.S.S. O'Malley**
Puri/ **L.S.S. O'Malley**
Gaya/ **L.S.S. O'Malley**
24 - Parganas/ **L.S.S. O'Malley**

BIHAR AND ORISSA DISTRICT GAZETTEERS

Patna/ **L.S.S. O'Malley**
Shahabad/ **L.S.S. O'Malley**
Monghyr/ **L.S.S. O'Malley**
Saran/ **L.S.S. O'Malley**

EASTERN BENGAL DISTRICT GAZETTEERS

Dacca/ **B.C. Allen**

BENGAL DISTRICT GAZETTEERS

24 - PARGANAS

BY

L.S.S. O'MALLEY

Indian Civil Service

LOGOS PRESS

New Delhi-110 002

ISBN-13: 978-81-7268-193-7
ISBN-10: 81-7268-193-3

2009

Printed and Published by

LOGOS PRESS

Building No. 4788-90, Street No. 23, Ansari Road, Darya Ganj
New Delhi-110 002 (India)

Phone : 23272187

Email : mail@logospressindia.com

Website : www.logospressindia.com

PREFACE

I desire to acknowledge, with gratitude, the assistance rendered by Babu Harendra Krishna Mitra, Head Clerk of the Bengal Census Office, in reading and checking the proofs. The Magistrate of the 24-Parganas has also been so kind as to have each chapter examined in his office.

L.S.S. O'MALLEY

CONTENTS

TABLE OF CONTENTS

CHAPTER-IV

Public Health

CHAPTER-V

Agriculture

CHAPTER-VI

Drainage and Embankments

CHAPTER-VII

Natural Calamities

CHAPTER-VIII

Rents, Wages and Prices

CHAPTER-IX

Industries, Manufactures and Trade

CHAPTER-X

Means of Communication

CHAPTER-XI

Land Revenue Administration

CHAPTER-XII

General Administration

CHAPTER-XIII

Local Self-Government

CHAPTER-XIV

Education

CHAPTER-XV

Gazetteer

GAZETTEER

OF THE

24-PARGANAS DISTRICT.

CHAPTER I.

PHYSICAL ASPECTS.

GENERAL DESCRIPTION. THE district of the 24-Parganas forms the south-western portion of the Presidency Division of Bengal, and lies between 21° 31′ and 22° 57′ north latitude and between 88° 2′ and 89° 6′ east longitude. It extends over 4,844 square miles, of which 2,941 square miles are part of the Sundarbans. It is more populous than either the North-West Frontier Province or the State of Baroda, its population, according to the census of 1911, being 2,434,104 : compared with European countries, it has about 400,000 more inhabitants than Wales. The district derives its name from the number of *parganas*, or fiscal divisions, comprised in the Zamīndāri of Calcutta, which was ceded to the East India Company in 1757 by Mīr Jafar, Nawāb Nāzim of Bengal. The head-quarters are situated at Alipore, a southern suburb of Calcutta, which, for the purposes of municipal administration, is under the jurisdiction of the Calcutta Corporation.

Boundaries. The district resembles an irregular parallelogram in shape, and is bounded on the north by the districts of Nadia and Jessore, on the east by the district of Khulna, on the south by the Bay of Bengal, and on the west by the river Hooghly, which, proceeding from north to south, separates it from the districts of Hooghly, Howrah and Midnapore.

Configuration. The 24-Parganas lies within the limits of the Gangetic delta, and its physical features are those common to deltaic land, for the country is flat, it is little raised above flood level, and the highest ground is that bordering the river channels. While these are the general conditions, the district naturally falls into two divisions with very different characteristics, viz, the northern inland tract, which is fairly well raised delta land of old formation and the low lying Sundarbans towards the seaboard on the south. The northern tract is a land of sluggish or stagnant rivers, whose beds are out of reach of the scour of the tides, and of inland depressions which will never now be filled, because the rivers, which should perform this office, are locked into their channels by the high banks of silt which they have deposited. The Sundarbans, on the other hand, are a network of tidal channels, rivers, creeks and islands. Some of these islands are mere swampy morasses, covered with low forest and scrubwood jungle, but those to the north, which are embanked, grow rich crops of rice. As one approaches the coast, the land gradually declines to an elevation which throughout many hundred square miles is scarcely raised above high-water mark. This seaboard area is a typical specimen of new deltaic formation. It exhibits the process of land-making in an unfinished state, and presents the last stage in the life of a great river—the stage in which it emerges through a region of half land, half water, almost imperceptibly, into the sea. It has been well described as "a sort of drowned land, broken up by swamps, intersected by a thousand river channels and maritime backwaters, but gradually dotted, as the traveller recedes from the seaboard, with clearings and patches of rice land."*

Industrial activity is concentrated in a narrow strip of foreshore along the Hooghly, extending from Budge-Budge (a few miles below Calcutta) to the northern limits of the district. This river frontage is densely populated, and is occupied by jute mills and crowded bazars, interspersed with Hindu temples and the gardens of country houses. Behind this strip of land the level drops, and the country is flat and uninteresting, until the eastern boundary is approached. Here the Jamuna river causes another rise in the surface, and the tract closely resembles Eastern Bengal; it is inhabited by sturdy Musalmān cultivators, who thrive on the abundant sugarcane and jute crops which they raise. In the north the monotonous level of the rice swamps is broken only by the clumps of palms and fruit trees in which the village homesteads nestle. The north-east of the district, where the land is higher than elsewhere, is studded with groves of date palm trees. In several places, more especially on the outskirts of villages, there are extensive plantations, the produce of which is boiled down into *gur* by the cultivators.

---

* W. W. Hunter, *Statistical Account of Bengal*, Vol. I, Preface.

Very different is the appearance of the Sundarbans, which is described as follows by Mr F E Pargiter*:—"The scenery in the Sundarbans possesses no beauty. The view even from a short distance is a wide stretch of low forest with an outline almost even and rarely broken by a tree rising above the dull expanse. In the forests, so far as I have seen them, there are few trees above 30 or 35 feet high, and few attain any considerable girth. This seems to be the result of the closeness with which they grow, and the poverty of the soil, which is impregnated with salt. But when a tree can get room enough to grow freely, it will attain a much greater size. The finest and largest trees I have seen have been almost invariably in places where the land had once been cleared; so that they had a good start before the jungle sprang up around them. There is little undergrowth in the forests, though here and there one may find cane-brakes and thickets of prickly scrub; and there is more of matted undergrowth and tropical luxuriance to be found in Sāgar Island than elsewhere. Few of the forest trees display a handsome bloom, as far as I have noticed at all times of the year except during the rains; the prettiest is a species of *hibiscus*, which grows freely along the banks of the streams, and bears large yellow flowers which turn to crimson as they droop. The only views that have some charm are to be found when drifting silently along with the tide, on a bright day in February and March, in the smaller streams in the Backergunge forest. The low *golpāta* palm with its immense leaves, the thickets of the *hibiscus* with its yellow and crimson blossoms, and clumps of the dark-green prickly *kewa* grow along the banks and overhang the stream, while above them the forest closes in with breaks of sunshine streaming through the foliage."

River system.

The rivers of the district were formerly distributaries of the Ganges, *i.e.*, they were the channels by which its waters were distributed and ultimately discharged into the sea; but the main current of that river has long since been deflected to the east, and their connection with its channel has been closed or silted up, so that they have ceased to be effective effluents. A large volume of water is, it is true, conveyed to the Hooghly, during the flood season, by the three Gangetic distributaries of which it is the product, viz., the Bhāgīrathi, the Jalangi and the Mātābhānga, while the Ichāmati carries flood water, to a limited extent, from the Mātābhānga to the north of the district. Other rivers, however, no longer receive a supply from the Ganges,

* *The Sundarbans,* Calcutta Review, Vol. LXXXIX, 1889.

and they cannot, as formerly, spread over the land and raise it by the deposit of silt. They now merely serve to receive local drainage and carry it off into tidal waters. The slope of the country being small, they have scarcely any current, and their channels in course of time become choked with aquatic vegetation. The principal rivers are the Hooghly, Bidyādhari, Piāli and Jamuna, but practically each river forms the centre of a minor system of interlacing distributaries of its own. Many change their names at different parts of their course, re-enter the parent channel, and then break away again or temporarily combine with other rivers until, approaching the sea, they are merged in the estuaries which pierce the Sundarbans. These estuaries are tidal waterways, with little or no current down them, and their water is as salt as the sea.

Hooghly. The Hooghly, which is the most westerly of the channels by which the waters of the Ganges enter the Bay of Bengal, marks the western boundary of the district. After receiving the Bāgher Khāl on the left bank, it flows in a southerly direction to Calcutta, below which it turns off first nearly due west, and then south-west as far as Achipur, from which point it flows almost due south, receiving the Dāmodar opposite Falta Point and the Rūpnārāyan opposite Hooghly Point. These great tributaries deflect the stream to the east for 8 miles, and have set up in it, just above the Rūpnārāyan, the dreaded moving shoals known as the James and Mary Sands. After Diamond Harbour the river resumes a southerly direction, until it debouches in the Bay of Bengal, its breadth at the point of junction being about 16 miles: its mouth is locally known as the Burha Mantreswar. Shortly before it falls into the sea it bifurcates, the main channel passing west and another channel east of Sāgar Island. This latter channel is called the Baratala River or Channel Creek, but is known locally as the Muriganga; it is fed by several subsidiary channels or creeks, and loses itself in the Bay east of Dhoblāt. On the right or west bank the main channel receives the Haldi river from the Midnapore district opposite Mud Point, on the north of Sāgar Island, and about 16 miles lower down the Rasūlpur joins it from the same district. In its course along the boundary of the district the Hooghly receives no important tributaries on its left bank, its only feeders being the Diamond Harbour and Khola Khāli creeks and the Falta, Nila and Harua Khāls, all small streams.

During the rainy season the spill streams from the Ganges and the Chota Nagpur tributaries of the Bhāgīrathi our down an enormous volume of water, which serves to scour out and maintain a deep channel. In the dry season, when there is no such influx, the river is largely fed by percolation, *i.e.*, by the underground infiltration of water into the deep trough which the river has scooped out for itself. In addition to these sources of supply, the Hooghly is a tidal river, the tides running up strongly, more especially during the dry season, when they provide water for navigation over the shoals; it is estimated that the tidal inflow during the four months of the dry season is more than double the total fresh-water discharge of the year. The greatest mean rise of tide takes place in March, April and May, and is about 16 feet; there is a declining range during the rainy season to a mean of 10 feet, and a minimum during freshets of 3½ feet.

Navigation is rendered difficult, not only by rapid currents, but also by shoals and shifting sandbanks. The James and Mary Sands, in particular, have an evil reputation for their peril to vessels making the passage of the Hooghly. They lie between Hooghly Point and Geonkhāli, and owe their formation to the Rūpnārāyan and Dāmodar. These rivers enter the Hooghly on its right bank at a short distance from each other, and, arresting the flow of its current by their combined discharge, cause it to deposit silt, which forms the shoal known by this name. The name itself is derived from the *Royal James and Mary*, a ship which was lost here in 1694. Since then many ships and steamers have been wrecked on the sands. Among others, the *Arcot* and *Mahratta*, steamers of the British India Steam Navigation Company, were lost in 1885, and the *City of Canterbury* in 1897, Above the James and Mary are other bars at Māyāpur. Rājpur, Falta and Naināu, and below them are the Upper Belari, Lower Belari, Haldia, Gābtala, Middleton and Gasper.

In spite of natural difficulties, the Hooghly is navigable by large liners, and is the gateway of the foreign trade of Bengal. Fears have from time to time been expressed that the river is deteriorating, and that it may eventually become inaccessible to maritime commerce. The records of the last century, however, show that the position of the Hooghly as a highway of commerce has not only been maintained but improved. How much has been done by engineering skill and systematic surveys may be realized by the following extract from a speech delivered towards the end of 1912 by Sir Frederick Dumayne, Vice-Chairman of the Commissioners for the Port of Calcutta:—

"The cry that the Hooghly is deteriorating is raised every now and then. Sixty years ago it was stated that the river threatened at no distant date to render access to Calcutta altogether impracticable for any but vessels of the smallest tonnage; that several vessels had no prospects of getting to sea for over a month, and that the serious impediments in the river were progressive and inevitable and beyond the reach of removal by any engineering skill. Ten years later there was a scare that the trade of Europe with Calcutta was in danger of immediate and prolonged suspension. The calamity which had overhung the city for years, and which Lord Dalhousie had, it was said, striven in vain to divert, and the fear of which had at intervals strained and baffled the ingenuity half the engineers of Bengal, was officially stated to be already and. Still, only with lessening fears, the same pessimistic valus are occasionally expressed. But what are the real facts? In 1830 pilots were prohibited from moving vessels of greater draft than 20 feet in any part of the river, even with the aid of tugs. In 1857 the permissible draft was only 22 feet. The average size of the largest vessels in the world was then about 2,500 tons burden, with a length of about 300 feet. The draft allowed has since been steadily increasing, until at the present time the river is navigated by vessels drawing up to 29 feet, of a length exceeding 500 feet, and carrying as much as 12,500 tons of cargo. The advent of steam, the great skill of the members of the Bengal Pilot Service, the elaboration of the system of surveying the river, were all factors in obtaining these results. Satisfactory as they were, further effort was demanded. Nature required assistance. The dredging of the river was undertaken in 1907. Hitherto only one dredger has been employed on this work, and her operations have been confined to the bars in upper reaches, with such success that these bars have never been in the good condition that they are in at the present time. It may be said that, by dredging, these bars have been brought under control. The dredging operations are now about to be extended, and contracts have been placed, in shipyards on the Clyde, for additional dredging plant that will cost £220,000, which include two very powerful dredgers, one of which will be employed on the improvement of the river within the port, and the other in dredging the bars in the lower reaches. The Commissioners confidently expect to obtain by dredging as good results in the lower as in the upper reaches, and ultimately to maintain throughout the year a clear channel from Calcutta to the sea for any vessel that can pass through the Suez Canal. The tonnage of vessels entering the port 50 years ago was 668,000 tons. Last year the tonnage was 6¾ million tons, or ten times as much. The growth has been most rapid in the last 15 years, when the increase was 3 million tons, or a yearly expansion of 200,000 tons. The value of the trade of the port of Calcutta from imports by all routes has reached 104½ millions sterling, and from exports 107½ millions, altogether 212 millions, an increase of 77½ millions in the last eight years."

The tide of the Hooghly is occasionally so strong that it gives rise to the phenomenon known as a bore. This is the name given to the headwave which is formed when an unusually high tide is checked by the narrowing of the river channel "The obstructed influx," writes Hunter in *The Indian Empire*, "no longer able to spread itself out, rises into a wall of waters which rushes onward at a rate nearly double that of a stage coach. Rennell stated that the Hooghly bore ran from Hooghly Point to Hooghly town, a distance of about 70 miles, in four hours." The height of the wave has often 5 feet or more, and old accounts testify to its danger. A writer in the Calcutta Review of 1859, for instance, states: "Upon the approach of this wave a distant murmur is heard, which turns into the cry *Bān! Bān! Bān!* from the mouths of thousands of people, boatmen, sailors and others, who are always on the look-out for this much dreaded wave. This cry is the signal for all sorts of craft to push out into the centre of the river, the only spot where the wave does not curl over and break. Should any boat or larger craft be caught in that portion of wave that breaks, instant destruction is inevitable. Numerous boats from the up-country provinces are lost every year from the crews being ignorant either of the existence of the bore, or from their not knowing the correct position to take up so as to meet it. Ships at anchor in Calcutta, though not exposed to the breaking portion of the wave, frequently part their cables when struck with the wave. Standing on the shore during the rapid rushing passage of the bore, it is a curious sight to see the lower portion of the river, or that nearest the sea, six or eight feet higher than the upper portion of the river, the tide rising that number of feet in an instant." The bore is not felt much till it enters the narrower reaches above Hooghly Point; and in recent years there has been no such high destructive wave as those described in the old accounts, presumably owing to the deepening of the river channel—for deep water robs the bore of its force, and it is chiefly to be feared by craft insecurely moored in shallow water or near a sand bank.

The present channel of the Hooghly is very different from that which the Ganges formerly followed. The original channel was identical with Tolly's Nullah from Kidderpore to Garia (8 miles south of Calcutta), from which point it ran to the sea in a south-easterly direction. Tradition has it that it emerged out of the Sundarbans at Kākdip, and then passed along the present Muriganga or Baratala river, after which it found a passage along a creek between Dhoblāt and Monosardip, and proceeded first in a westerly and then in a southerly direction until it fell into the Bay of Bengal at Ganga Sāgar. It is still traceable as far south as the Jaynagar thana, being known variously as Adi Ganga (*i. e.*, the original Ganges), Burha Ganga (or the Old Ganges) and Ganga Nullah. It still retains its sanctity in the eyes of Hindus, who burn their dead along its banks, whereas the present Hooghly river below Tolly's Nullah is not considered a sacred stream. The old channel may still be made out at various places, such as Bāruipur, Dakshin Bārāset, Jaynagar and Rājpur in the Sadar subdivision and Multi and Hānsghar in the Diamond Harbour subdivision. Even in places where it is entirely silted up, tradition points to old tanks, still called Ganga, as having been excavated along its course while it was an active stream. Such tanks may be found at Jaynagar and Vishnupur (Bistupur) and Khāri in the Mathurāpur thāna: that at the place last named has a special sanctity, and is known as Chakratīrtha or Chakraghāta. Their water is accounted as sacred as that of the present Ganges, and Hindus bathe in them and burn their dead on their banks with the same assurance of spiritual benefit.

Bidyādhari and Piāli. The Bidyādhari is a tidal river with a very circuitous course. Beginning in the Sundarbans, it flows north-east past Hārua, where it is known as the Hārua Gāng, and then bends westwards, and is joined by the Nona Khāl. After this, it flows south-westwards to the junction of the Baliaghāta Canal and Tolly's Nullah, and thence south-east to Mātla or Canning, where it is joined by the Karatoya and Athārabānka rivers. The united stream forms the Mātla river, which flows south to the sea and is navigable by river steamers up to Canning.

The Piāli leaves the Bidyādhari 9 miles below Bāmānghāta and flows south and south-west till it joins the Mātla river about 20 miles below Canning. It is bridged by the Canning branch of the Eastern Bengal State Railway.

The portion of the Bidyādhari near Calcutta, which at present serves as an outfall channel for the storm-water and sewage of the city, has for some years past been silting up at a rapidly increasing rate. The acceleration of the silting process is attributed mainly to works in connection with local fisheries and to the reclamation of portions of the Salt Water Lakes for rice cultivation, the effect being to decrease the spill of water from the river over the adjoining land and, consequently, to increase the deposit of silt in the river bed. Other contributory causes have been the construction of the Dhāpa lock, the closing of tributaries in each of which the tide used to flow and ebb freely, and the canalization of the Bhāngar Khāl. Observations taken between 1904 and 1912 show that, a mile below Bāmanghāta, the bed of the river has risen nearly 25 feet in eight years, while in the section immediately below Bāmanghāta lock the cross-sectional area has been reduced from 7,700 square feet to 3,870 feet, giving a mean rate of contraction of 480 square feet per annum. "It is," says Mr. O. C. Lees, C.S.I., Special Officer, Hooghly-Bidyādhari Canal Enquiry, "practically certain that this mean rate taken over eight years does not give the measure of the deterioration in progress to-day. But, even if it be assumed that the rate has increased only to 500 square feet per annum, then it is apparent that the bed of the river five years hence will have risen above the level of ordinary high water of neap tides." He concludes that the Bidyādhari has a very short remaining lease of life, and that in six years' time it will be useless as an outfall channel for the sewage of Calcutta unless remedial measures are taken. In his opinion, the best and indeed, the only practicable way of dealing with the situation, is to canalize the channel of the Bidyādhari from Dhāpa to the offtake of the Piāli river, 9 miles below Bāmanghāta, and the channel of the Piāli river from its head to its outfall into the Mātla river. A dam would be thrown across the Bidyādhari below the head of the Piāli, while the latter would be dammed at its mouth, a lock being constructed to admit boats and a large drainage sluice to discharge the drainage from the areas of which the upper Bidyādhari and the Piāli are the natural drainage channels.

Jamuna or Ichāmati. The Jamuna or Jabuna enters the district from Nadia at Baliāni, and flows south-east to Tibi, where it is joined by the Ichāmati. From Tibi it flows in a very irregular south-easterly direction past Bāduria, Basīrhāt, Tāki, Srīpur and Debhāta to Basantpur. In this part of its course the combined stream is usually known as the Ichāmati. From Basantpur it flows south-east through the Sundarbans in Khulna to the sea, where it falls into the Raimangal estuary. The canals, or navigable channels, which run eastward from Calcutta, join this river at Hasanābād.

Kālindi. The Kālindi branches off from the Jamuna at Basantpur and forms the eastern boundary of the district down to the sea, where it merges in the Raimangal estuary.

Kalāgāchia. The Kalāgāchia river is formed by the combination of several Sundarbans *gāngs* or water channels in the Hasanābād thana, and flows in a south-easterly direction. It is connected with the Mātla by the Athārabānka river.

Other rivers. The south of the Sundarbans is traversed by numerous tidal rivers, some of them of great size, formed by the junction of smaller water-courses and branches thrown off by other rivers, all having a general southerly course towards the sea. The most important from west to east are—(1) In Kulpi thana, the Baratala river or Channel Creek, an offshoot of the Hooghly; (2) between the Kulpi and Mathurāpur thanas the Sabtarmukhi river; (3) in Jaynagar thana, the Thākurān, which afterwards becomes the Jamira river and forms the boundary between the Mathurāpur and Jaynagar thanas; (4) also in Jaynagar thana, the Kaikalmāri, an offshoot of the Mātla, which it rejoins a little further south in Hasanābād thana, (5) the Guāsuba river, which reaches the sea under the same name; and (6) the Sursa river, which lower down becomes the Hariābhānga, and under that name joins the Raimangal estuary.

ESTUARIES. The chief estuaries, or arms of the sea, from west to east, are the Baratala or Channel Creek, the Sabtarmukhi, the Jamira, the Mātla, the Guāsuba and the Raimangal, the last including the mouths of the Hariābhānga and Kālindi.

ISLANDS. These estuaries are interspersed and separated by several large islands, of which the chief, from west to east, are as follows:—(1) Sāgar Island between the Hooghly and Channel Creek. This, the largest and most important of the islands, is itself cut up into many smaller islands by cross-streams: (2) Mecklenberg Island, recently renamed Fraserganj, west of the Sabtarmukhi; (3) Lothian Island in the mouth of the Sabtarmukhi; (4) Bulcherry (Balchāri) Island between the Jamira and Mātla; (5) Halliday Island in the Mātla; (6) Dalhousie Island between the Mātla and Guāsuba; (7) Bāngaduni Island at the mouth of the Guāsuba.

At the south-western extremity of Sāgar Island stands the Sāgar lighthouse; a large fair, known as the Ganga Sāgar *mela*, is held on its southern face every year in January. There is also a beacon at Dalhousie Point, the south-eastern corner of Dalhousie Island on the east of the mouth of the Mātla.

Khāls. Between the large estuaries and rivers are innumerable streams and water courses, called *khāls*, forming a perfect network of channels, and ending ultimately in little channels that serve to draw off the water from each block of land. For each block is formed like a saucer, with high ground along the bank of the *khāls* surounding it, and with one or more depressions in the middle, according to their size. The water collects in the depressions, and is drained off by the little *khāls* into the larger *khāls*, and ultimately into the rivers; conversely, when the water swells in the rivers, it floods the country through the same channels. Many of the *khāls* connect two large ones, and consequently the tide flows into them through both ends; such *khāls* are called *do-āniya khāls*. They are very useful as affording communication between the larger *khāls*, but have one serious defect, that they are liable to silt up at the point where the two tides meet, for the water is always still there.

"In the Sundarbans," writes Mr. Pargiter* "there is generally a belt of jungle along the banks of such *khāls*, for no cultivation can be attempted outside the embankments, where the salt water can reach. These *khāls*, therefore, are closely shut in by jungle, and the condition of the water in them increases their insalubrity. If the *khāl* is open, the tide flows in and out, and leaves, except at high tide, a bank of mud, which is as much as ten or twelve feet high in places near the sea. If the mouth of the *khāl* has been dammed, the water is necessarily stagnant and unwholesome. The people cannot escape these unhealthy conditions, for roads are unknown in the Sundarbans, and the waterways are the only means of communication with outside places; while the embankments are the chief means of communication within the blocks of land. Wholesome fresh water can scarcely be had anywhere, and even when ordinary fresh water is not obtainable, the people will use water that is slightly brackish, without apparently any deleterious consequences. Along the Hooghly, the land is high, but east of it, as far as the river Kālindi, the land is lower, and is liable to be submerged, and immense embankments are indispensable to its reclamation."

The 24-Parganas, like other deltaic districts, is studded with large marshes and swamps (*bils*) situated between the elevated tracts which mark the courses of the rivers. Their presence is the natural result of the configuration of the country. The river banks silt up till they become the highest levels, from which the ground gradually slopes downwards, forming a depressed tract between each set of two rivers. The depressed portions so constituted are natural basins, in which water collects and from which it has no exit. These marshes are chiefly situated in the east of the district, and the principal are as follows:— **MARSHES.**

(1) The Dhāpa or Salt Water Lake between the Hooghly and Bidyādhari is a low basin which commences about five miles

* *The Sundarbans*, Calcutta Review, Vol. LXXXIX, 1889.

east of Calcutta, and has an area of about thirty square miles A part of the lake is in course of reclamation, as it is slowly filling up with silt deposited by tidal channels from the Bidyādhari, and also by the deposit of the street refuse of Calcutta. (2) The Kulgāchi Bil, situated to the west of Bālinda, in the centre of the insular portion surounded by the Bidyādharī river and the Kātākhāl. (3) The Bariti Bil, half way between Bālinda and Basirhāt, between the Bidyādhari and Jamuna rivers. (4) Bayra Bil, the largest in the district, comprising an area of forty square miles, the greater part of which is covered with reed jungle; it is situated east of the Jamuna, in the north-east of the district, and is being gradually drained and reclaimed. (5) and (6) The Balli and Dāntbhāngā Bils, the former containing ten and the latter twelve square miles, are also situated east of the Jamuna, to the north of the Bayra Bil. (7) The Bartti Bil, east of Shāmnagar, a station on the Eastern Bengal State Railway. (8) The Dholkhera Bil, south-east of Bārāset. (9) The Ghāzalmāri Bil. (10) and (11) The Nagarghāta and Khālishkhāli Bils, east of the Bayra Bil.

Large areas of marsh land in the Sundarbans have been reclaimed and brought under tillage by means of embankments raised to keep out salt and brackish water. Some of the large marshes in the heart of the district have also been drained and rendered fit for cultivation by the construction of embankments and sluices; but as the basins served by these sluices are dependent on each other, it follows that whenever there is abnormally heavy rainfall, areas which are properly provided with sluices are liable to suffer from the overflow of adjoining tracts, and the large central basins, which have no proper facilities of their own for the discharge of their surplus water, add to the volume contained in the surrounding basins. The most notable drainage work effected of late years is that known as the Magra Hāt drainage scheme, of which an account will be given in a subsequent chapter. Here it will suffice to say that this scheme provides for the drainage of an area of nearly 300 square miles in the south of the district, which used to be so waterlogged as to be almost a morass. The central portion, indeed, was described by the Executive Engineer, in 1879, as little better than a permanent jheel. "In the dry weather the tides overspead the low land with salt or brackish water, and the rains inundate and destroy the crops. From want of drainage and protection, the productiveness of the locality is only a fraction of what it should be, and the inhabitants, though they may be supposed to be inured to their semi-amphibious condition by a long course of preparation, resulting in the survival of the fittest, are affected similarly to those living in the vicinity of the permanent *bils*. Fever is constantly present in every village, and other classes of sickness find a congenial home." Though the works designed for the drainage of this tract have only recently been completed, the area under cultivation has already been extended largely, and the tract has had a substantial increase of population.

GEOLOGY.

The soil of the district is composed of alluvium, which is of great depth. A boring at Calcutta reached a depth of 481 feet without signs of either a rocky bottom or marine beds, while fragments of fresh water shells were found as low as 380 feet below the surface. The most remarkable fact connected with the geological formation is that there are reasons for believing that there has been some subsidence of the country. This is apparent from the discoveries made 50 years ago by Colonel Gastrell,* who wrote :—"What maximum height the Sundarbans may have ever formerly attained above the mean tide level is utterly unknown; that they ever were much higher than at present is, I think, more than doubtful. But that a general subsidence has operated over the whole extent of the Sundarbans, if not of the entire delta, is, I think, quite clear from the result of examination of cuttings or sections made in various parts, where tanks were being excavated. At Khulna, about 12 miles north of the nearest Sundarbans lot, at a depth of eighteen feet below the present surface of the ground, and parallel to it, the remains of an old forest were found, consisting entirely of *sundri* trees of various sizes, with their roots and lower portions of the trunks exactly as they must have existed in former days, when all was fresh and green above them; whilst alongside them lay the upper portions of the trunks, broken off and embedded in a thick stratum of old half-decomposed vegetable mould, nineteen inches in depth, from which, when first exposed, leaves, grasses and ferns could readily be separated and detached. Below this were other thinner strata of clays and vegetable mould corresponding to the Calcutta peat, whilst above was a stratum of argillaceous sand passing into stiff blue clay containing numerous shells. One of the trees was found projecting far into the upper stratum of blue clay. Many of the trees were quite decomposed, whilst in others the woody fibre was nearly perfect."

A writer in the *Calcutta Review*† of 1859 has given an account of a similar discovery of *sundri* trees at Canning. "That forests

* Revenue Survey Report on the districts of Jessore, Faridpur and Backergunge.

† *The Gangetic Delta*, Calcutta Review, 1859.

now lie under the Sundarbans we have seen with our own eyes. In excavating a tank at the new town of Canning, at the head of the Mătla, large *sundri* trees were found standing as they grew, no portions of their stems appearing above ground; their numbers may be imagined when we state that in a small tank, only thirty yards across, about forty trees were exhumed, ten feet below the surface of the country, their timber undecayed, showing that no very great period of time has passed over their submergence. If the present level of their roots could suddenly become the level of the country, the whole Sundarbans would be under water. At a lower level than these trees, beds of a peaty mass, composed of decayed and charred wood, are pierced in Calcutta, Hooghly, Dum-Dum and elsewhere, at a depth varying from eight to eighty feet."

That this subsidence of the surface of the ground is not confined to the Sundarbans, seems to be confirmed by the fact that stumps of trees have also been found at Sealdah in Calcutta, at various levels down to a depth of thirty feet, or ten feet below the peat. These trees also were pronounced by the Superintendent of the Botanical Gardens to be *sundri*, a tree of which the range is restricted to from two to about ten feet below high-water mark. It appears, therefore, that the deltaic tract stretching from Khulna to Calcutta must at some time have undergone a subsidence. The following evidence in support of this view may be quoted from the *Manual of the Geology of India*, by R. D. Oldham (1893):—

"The peat bed is found in all excavations round Calcutta, at a depth varying from about twenty to about thirty feet, and the same stratum appears to extend over a large area in the neighbouring country. A peaty layer has been noticed at Port Canning, thirty-five miles to the south-east, and at Khulna, eighty miles east by north, always at such a depth below the present surface as to be some feet beneath the present mean tide level. In many of the cases noticed, roots of the *sundri* tree were found in the peaty stratum. This tree grows a little above ordinary high-water mark, in ground liable to flooding, so that in every instance of roots occurring below the mean tide level, there is conclusive evidence of depression. This evidence is confirmed by the occurrence of pebbles, for it is extremely improbable that coarse gravel should have been deposited in water eighty fathoms deep, and large fragments could not have been brought to their present position unless the streams which now traverse the country had a greater fall formerly, or unless, which is perhaps more probable, rocky hills existed which have now been covered up by alluvial deposits. The coarse gravels and sands, which form so considerable a proportion of the beds traversed, can scarcely be deltaic accumulations, and it is therefore probable that, when they were formed, the present site of Calcutta was near the margin of the alluvial plain, and it is quite possible that a portion of the Bay of Bengal was dry land."

As regards the cause of this subsidence various theories have been propounded. The writer in the *Calcutta Review* already quoted ascribes it to the weight of the superincumbent earth and forest. "If," he says, "we consider the unsubstantial nature of the foundation of the Sundarbans, which, at a distance of only 120 feet from the surface, consists of a bed of semi-fluid mud, 40 feet in thickness, and then remember the terrific convulsions that have at different periods shaken the delta to its deepest foundations, we must not be surprised to find that the liquid mass, unable to support the superincumbent weight, has repeatedly bulged out seaward, reducing the level of the delta submerging whole forests, together with their fauna and flora."* Colonel Gastrell, again, considered that, though the general depression might have been caused partially in this way, it was more probable that it was caused suddenly during some great earthquake. "The fact of all the trees being, as a rule, broken off short and none being found standing at Khulna or Sealdah, might, in that case, be accounted for by the enormous wave that, in such a subsidence, would have rolled in from the Bay over the Sundarbans, destroying all in its path. Or, supposing the subsidence not to have been general over the whole tract at first, and only sufficient to have submerged the roots below low-tide-level, and so killed them, all would have dried up as they stood, and succumbed to some one of the cyclones that must have subsequently swept over the tract. The latter assumption seems likely, because, whilst at Khulna and Sealdah the trees were all broken short off close to the ground, at Mātla, which is situated between these places, they are said to have been found intact and unbroken, which could not have been the case had a great wave, caused by the sudden subsidence of the country, swept in from the Bay over the sinking forests: in that case, Khulna, Mātla and Sealdah, supposing the submergence general, would have been exposed and suffered alike, and the trees would have been equally decomposed in all places. The fact of their not being so seems to clearly show that the subsidence at Khulna was prior to that at Mātla, as in the first place the trees were mostly decomposed, while in the

* *The Gangetic Delta,* Calcutta Review, 1859.

second they were not so. All is pure conjecture, however, and the causes may have been very different."

The process by which the present alluvial surface was formed has been well described by Dr. Thomas Oldham, in an article published in the *Proceedings of the Asiatic Society of Bengal* for 1870.

"I suppose no one will hesitate to acknowledge that the whole of the country, including the Sundarbans proper, lying between the Hooghly on the west and the Meghna on the east, is only the delta caused by the deposition of the debris carried down by the rivers Ganges and Brahmaputra and their tributaries. It is also equally well known that in such flats the streams are constantly altering their courses, eating away on one bank and depositing on the other, until the channel in which they formerly flowed becomes choked up, and the water is compelled to seek another course. It is also certain that, in this peculiar delta, the general course of the main waters of the Ganges has gradually tracked from the west towards the east, until, of late years, the larger body of the waters of the Ganges has united with those of the Brahmaputra, and have together proceeded to the sea as the Meghna. Every stream, whether large or small, flowing through such a flat, tends to raise its own bed or channel by the deposition of the silt and sand it holds suspended in its waters, and by this gradual deposition the channel bed of the stream is raised above the actual level of the adjoining flats. It is impossible to suppose a river continuing to flow along the top of a raised bank, if not compelled to do so by artificial means, and the consequence of this filling in and raising of its bed is that, at the first opportunity, the stream necessarily abandons its original course, and seeks a new channel in the lower ground adjoining, until, after successive changes, it has gradually wandered over the whole flat and raised the entire surface to the same general level. The same process is then repeated, new channels are cut out, and new deposits formed.

"Bearing these admitted principles in mind, look to the delta of the Ganges and Brahmaputra. The Ganges river, emerging from its upper levels round the Rājmahāl Hills, and prevented by their solid rocky barrier from cutting further to the west, sought its channel in the lower ground adjoining, and originally the main body of its waters flowed along the general course now indicated by the Bhāgīrathi and Hooghly. But, gradually filling up this channel, it was again compelled to seek a new course in the lower, because as yet comparatively unfilled-in, ground lying to the east. And, the same process being repeated, it wandered successively from the rocky western limit of the delta-flat towards the eastern. If this progress eastwards was allowed to be sufficiently slow to admit of the gradual filling in of the country adjoining, the delta was formed continuously up to the same general level, and the larger streams or channels, passing through this flat to the sea, became unavoidably diminished in size and in the quantity and force of the water they carried, the main body passing round further to the east, and having its course in the channels successively formed there."

The following account of the botany of the district was contributed for the revised edition of the *Imperial Gazetteer* by Lieut.-Col. Sir D. Prain, C.M.G., C.I.E., the present Director of the Royal Botanic Gardens, Kew. BOTANY.

The stretches of low-lying land under rice cultivation afford a foothold for numerous marsh species, like *Sphenoclea*, *Hydrolea*, *Ammania*, *Enhydra*, *Wedelia*, *Hygrophila* and *Sesbania*, while the numerous ponds and ditches are filled with submerged and floating water plants, like *Utricularia*, *Pistia*, *Hydrilla*, *Lagarosiphos*, *Ceratophyllum*, *Ottelia* and *Vallisneria*. Remarkable among these for its rarity, and interesting on account of its distribution to Europe on the one hand, and Australia on the other, is the floating *Droserad Aldrovanda*. The edges of sluggish creeks are lined with large sedges and bulrushes; the banks of rivers have a hedgelike scrub jungle of *Derris*, *Dalbergia* and *Cæsalpina*, with climbing *Ipomœas*, *Argyreias* and *Menisperms*, and a few trees like *Pongamia glabra*, *Barringtonia acutangula* and *Thespesia populnea*. The sides of embankments and village sites, where not occupied by habitations, are densely covered with village shrubberies of semi-spontaneous species, like *Odina*, *Zizyphus*, *Acacia*, *Glycosmis* and *Trema*, often interspersed with clumps of planted bamboos and groves of *Areca*, *Moringa*, *Mangifera* and *Anona*. Waysides and waste places are filled with grasses and weeds, usually of little intrinsic interest, but often striking because of their distribution. A very large proportion of the species of this class to be met with in the district has been inadvertently introduced by human agency; besides weeds that are indigenous in other parts of India, these include European or African species like *Senebrera pinnatifida*, and *Xanthium spesidsum*, and especially American species like *Ageratum conyzoides*, *Scoparia dulcis*, *Wissadula rostrata*, *Evolvulus nummulasifolius*, *Pepromia pellucida*, *Malachra capitata*, *Herpestris chamædryoides*. *Croton sparsiflorus*, and many others, which not only hold their own with, but spread more plentifully than, similar weeds of truly Indian origin.

WILD ANIMALS.

The following account of the wild animals, game birds and reptiles met with in the 24-Parganas is given by Mr. F. B. Bradley-Birt, I.C.S.:

"The following wild animals are found in this district:—*Order Ungulata.*—Barking deer (*Cervulus aureus*), spotted deer (*Axis maculatus*), hog deer (*Axis porcinus*) and wild pig (*Sus Indicus*). *Order Carnivora.*—Tiger, panther, fishing cat, civet cat, palm civet, mongoose, monkey, jackal, fox, and others. The more interesting are described below:—

"The *chital* (*Cervus axis*) is commonly known as the spotted deer of India. It is reddish brown in colour, with distinct white spots along its back and sides. The colour varies, however, from a light reddish brown, in young stags, almost to black in old males. The shading is darkest about the back, gradually fading as it approaches the ventral aspect, which is white. One instance of albinism in a stag has been met with. Each horn has three distinct points, but the short snags just above the brow antlers, found in stags in other parts of India, are absent in the Sundarbans variety. The ordinary length of horn is about 25 inches, with a thickness of 3½ inches. These are shed very irregularly, antlers in velvet being met with at all times of the year. This deer is common in the jungles of the Sundarbans, and is found in fairly large herds along the sea coast, where long glades of short green grass are to be met with. It causes great destruction to paddy crops in new clearances, leaving cover early in the afternoon, and approaching, as night comes on, quite close to the hamlets of villagers. Its cry is a short high-pitched kind of bark, generally used as a signal or alarm, but also when calling to one another, the intensity of pitch and sharpness in a cry of alarm being greater than in an ordinary call. Ordinarily it drinks the salt water of the rivers and creeks, though it shows its preference for fresh water as soon as tanks are dug in a new settlement. During the rutting season, stags have been observed in long combats, pushing each other, with antlers interlocked, up and down a clear bit of jungle, and uttering guttural sounds. The *chital* is readily tamed, and tame ones are used by *shikāris* as decoys when shooting. Barking deer and hog deer are not uncommon, but, being very shy, are seldom seen along the banks of streams. They are found in the reserved forests and uncultivated parts of the northern side of the Sundarbans. Little need be said of the Bengal monkey. He is harmless, and is common in all tracts which are still under jungle.

"The wild pig abounds in the jungles of the Sundarbans, and is fairly common over the rest of the district, being met with wherever clumps of bamboos or undergrowth on waste lands afford it cover. The young of the animal is dark brown, with longitudinal stripes of a lighter shade. These stripes disappear after a few months, and the animal then puts on a black coat. The boar is armed with long tusks, which, in a veteran of solitary habits, may measure over 9 inches. Wild pigs go about in sounders of from four to ten individuals, according to the nature of the ground which affords them cover. They are particularly destructive to paddy in ear. In walking through standing crops, the stalks coming between their forelegs are pressed down, and the grain thus brought to their mouths. The animals go slowly through the fields, leaving behind them long lines of destroyed plants. Except in the northern part of the district, it is quite impossible to spear pig, as the country is not suitable for riding. The animals have to be shot down. This is, however, a poor form of sport, as pigs are not particularly alert, or shy of human beings, and can be easily got at. For the former reason they furnish the ordinary food of the tiger.

"The *bāgh* or tiger (*Felis tigris*) is fairly common in the jungles of the Sundarbans. In fact, it is impossible to land anywhere in that tract without coming across the pugs, or foot-prints, of that animal. But inferences drawn from such marks, as to their numbers, would be erroneous, unless it is remembered that tigers are great wanderers, covering large distances in search of prey, and readily swimming across the rivers and *khāls* which intersect the jungles on all sides. As an instance of their swimming capacity may be mentioned the fact that a stray tiger was not long ago found at the mouth of the Rasūlpur river, in the Midnapore district, which had swum across from Sāgar Island, the breadth of the river between being about eight miles. The Sundarbans tiger is not so large as his congener along the Tarai, or in the Central India jungles, the average length of males being 9 feet and or females 8 feet 6 inches. In their marking these animals vary greatly. The young tigers are handsomer than the old ones; their colouring is more vivid, and the stripes darker and closer together. A curious adaptation to environment is seen in the tigers frequenting the sand dunes of the sea face. These sand dunes are covered with tall, brown spear-grass, and immediately behind lie the glades of green herbage, on which deer feed and pigs come out A tiger, in such surroundings, would be rendered conspicuous by his stripes, so the sand dune tiger has almost lost his stripes. The last specimen examined by the writer had barely half a dozen thin dark lines, mostly about the shoulder, and the coat of the animal was of a tawny orange colour, well adapted to the grass or sand dunes. It was not an old tiger. The age of tigers can be told by the ridge of bone that runs along the top of the brain-pan, which becomes more prominent as age increases, or by the canine teeth. In the adult they are hollow throughout, and the base is open. With age the base gets closed, and the tooth, so far as it shows from the gum, is solid. Tigers are not gregarious in habit: more than two are seldom seen together, and though these may frequent the same haunt in the day, they almost invariably separate at night while searching for prey. They are seldom seen in these parts during the day. These animals are extremely suspicious, and with their senses well developed. The writer once tied up cattle in the glades along the sea coast, and, though tigers were in evidence on the spot for three nights in succession, they failed to attack such an unusual form of prey. Once, however, beef is tasted, they soon become confirmed cattle destroyers, and cause great loss to the villagers of the *ābāds*. Tigers in the Sundarbans very often carry away wood-cutters and others who frequent the reserved forests for wood or other forest produce, and some man-eaters have been known to stop work in new clearances by the frequent toll levied on human life. The writer has come across a well authenticated instance where such a man-eater charged into a line of some six or eight men, working along a *bund*, at about 8 or 9 A. M., and carried off a man from their midst. They are numerous in the Khulna portion of the Sundarbans, where they cause much loss of life. They formerly gave trouble in the Backergunge and 24-Parganas tracts of the Sundarbans; but, with the steady extension of cultivation, their number has decreased, and they have been forced to move southwards towards the Bay, where they live chiefly on wild pig and deer. Tigers breed during the months from December to June, the period of gestation being from 14 to 15 weeks. No elephants can be used in the jungles of this district, and, owing to the dense growth of the wild date and cane, it is impossible to beat. The only method of killing tigers is by sitting up over kills or bait. Like other animals in the Sundarbans, tigers drink the salt water of the rivers and creeks, so that, as a rule, there are no water-holes to sit over. The Government reward for their destruction is Rs 50 on the forest reserves, and Rs. 25 elsewhere.

"*Chitta bāgh* or panther (*Felis pardus*). The animals known to sportsmen as leopard and panther are really of the same species, the difference being merely that of size and colour, which varies from melanism to albinism, though the latter is extremely rare. The panther is not found in the Sundarbans jungles, nor is it anything but rare in the northern parts of the district. An occasional animal is heard of from time to time as frequenting some jungle or waste lands, but as soon as it takes to killing goats and young cattle its doom is assured. Panthers have been known to develop into extremely dangerous man-eaters, their power of climbing giving them an advantage over tigers, but no such instance has been reported in this district of late years.

"The fishing cat (*Felis viverrina*) is found in the Sundarbans, though it is a very rare animal. One specimen was captured by the writer and sent to the Zoological Gardens. It was a full-grown specimen, beautifully marked with the rosettes of a leopard, of which it was a picture in miniature. It had taken refuge in a tree during the high water of a spring tide. Though it could have swum across a fairly broad stream and escaped, it did not do so, but kept going short distances from tree to tree. The civet cat (*Viverra zibetha*) and palm civet (*Paradoxurus*) are found in the district, the latter being rare; but for a description of these animals the reader is referred to standard works on mammalia. The jungle cat (*Felis chaus*) and civet cat are found in the vicinity of inhabited and reclaimed tracts. The *ud* or otter (*Lutra nair*) is found in small colonies in parts of the Sundarbans. The animal is tamed by fishermen, and trained to drive fish into their nets.

"In the Sundarbans there are no game birds, except the jungle fowl. Owing to the absence of *jhils*, no ducks are found, beyond small flights of common teal on some of the rivers. Snipe are not very plentiful. Sāgar Island, however, is resorted to by geese (greylag), which feed on the paddy fields at night. The northern part of the district possesses some *jhils* and *chars* which are visited in the winter by the usual migratory ducks—common mallard, pochard, shoveller, gadwall, teal, pintail, widgeon. But the country here is well populated, and many guns are possessed by *shikāris* and villagers, so that ducks are too often disturbed to be plentiful. Nor is their stay very long. Snipe (pintail, common and painted) are found during the winter on paddy land. Golden plover are also met with when the paddy has been cut. Partridge and quail (button) are scarce. The only other game bird is the curlew, which is found on the banks of all streams. He is shy to a degree, and large flocks are often seen flying off on the least sign of danger. When shot during the cold weather, the flesh of the curlew of the Sundarbans is very delicious, but after March it tastes fishy, and is sometimes uneatable. The younger birds are best for the table. Of other birds, the common vulture (black and brown), kites, hawks *mainas*, golden orioles, sandpipers, egrets, plovers, waders, pelicans, GAME BIRDS.

herons, owls, doves, crows, woodpeckers, jays and other small birds are seen.

FISH.

"The fish of these waters belong mostly to the carp and dog-fish families. They are plentiful and found at all times of the year. The growth of Calcutta has created such a demand for this article of diet, that in the neighbourhood of Dhāpa, Bāmanghāta, Garia and other parts, paddy land has been converted into fisheries by breaking the *bunds* in places, and allowing the tide water to flow over the land along various small channels, at the mouths of which nets and traps are laid. Fishing also goes on largely amongst the creeks of the Sundarbans, each day's catch being put into large bamboo cylinders which, when full, are towed up to the nearest marts. The varieties commonly found in the Sundarbans are the *bhekti*, *hilsa*, *tapsi*, *bhāngan*, *kān-māgur*, *rekha*, *rucha*, *chitra* or *bisatara*, *pārse*, *tengra*, *bānspāta*, *ditnia*, *māgur* and *kai*. Crabs and shrimps are common. Good business is done in shrimps, which are dried and despatched to Chittagong and Burma. Sharks are common in these parts, and are caught in nets for the sake of their fins, with which the Maghs and Arakanese do good business, and obtain large profits.

REPTILES.

"The *kumbhir*, or marsh crocodile (*Crocodilus palustris*), is found in tanks, *jhils* and the upper reaches of all the rivers in the district. The larger animals are by no means averse to attacking man, though this crocodile generally lives on fish and the common domestic animals, which it either catches itself, or the carcasses of which are thrown into the river by the inhabitants. It is a cannibal, and will feed on the remains of its own kind. Another species, the estuarine crocodile (*C. porosus*), inhabits the lower reaches of the rivers right up to the sea face. It is emphatically the man-eating crocodile of India. It differs from *Crocodilus palustris* in having a narrower head, with two ridges on the skull converging towards the snout, and, in the adult, four instead of five teeth in the upper jaw, counting from the median line to the notch. Both species are much given to basking on the sand spits and banks which are left uncovered at low tide, where, especially during the winter, they will be found lying with their mouths wide open. The writer has not been able to discover why they should open their mouths at that season, and not at other times. The *Crocodilus porosus* grows to great lengths: a specimen in the British Museum is recorded to be 33 feet. It breeds in certain favourite rivers; the eggs are laid amongst a mounds of leaves, the fermentation of which hatches the young. The Fuldobi creek, for instance, will be found in January and February full of young crocodiles. As the animals frequent the same spots, year after year, they can be waited for and shot. The cry of a crocodile is a long drawn howl, which the writer has had the opportunity to hear on two occasions. The snub-nosed crocodile (incorrectly called an alligator) may be seen in large numbers along banks of streams, especially during the cold weather. He is known to overturn boats for the purpose of killing human beings At times they are so daring that they enter homesteads at night and carry off the first person they can seize. They also enter fields during the day, and kill and carry away cattle. They are most ferocious, and cause great loss of life. They are often harpooned by village people, who go after them in boats for the purpose, and not unfrequently bring them in amidst the rejoicings of the entire village ".

During the cold weather, from November to January, there is an almost entire absence of cloud and rainfall. The total mean rainfall for these three months is very little over one inch. The mean temperature falls from 72° in November to 65° in December and January, but humidity continues high. A feature of the cold season is the occasional occurrence of low-lying fogs, which dissipate with the rising sun. In February the temperature begins to rise, the mean for the month being 69°. Southerly winds become more frequent with the advancing year, and there is a period of transition characterized by occasional thunderstorms, accompanied by rainfall. The average amounts to 1 inch in February and 1·7 inches in April. The mean maximum temperature is 96° in April, and there is a decline of 1° in May. Night temperature increases slowly, and the highest monthly average is not reached till June and July, when it is 79°. The mean temperature for the whole year is 78°. CLIMATE.

In May, monsoon weather is occasionally experienced when cyclonic storms occur near the head of the Bay. These storms bring heavy rainfall, and the average consequently rises from 1½ inches in April to 5½ inches in May. In some years the fall is far heavier, *e.g.*, in 1893 the total fall at Calcutta during May was 25 inches, in 1878 it was 15 inches, and in 1865 it was 16 inches. With the commencement of the south-west monsoon, which generally occurs in the latter half of June, but sometimes is deferred till the beginning of July, humidity increases to 90 per cent. of saturation, while heavy cloud is continuous and rainfall of daily occurrence. The average rainfall is 11 inches in June, 13 inches in July and August and 10 inches in September. The mean temperature slowly diminishes from 85° in June to 80° in October. During the latter half of September, and throughout October, cloudy weather alternates with bright sunshine, the bright periods lengthening until they merge into the continuous fine weather of the cold season.

One of the most remarkable cases of heavy rainfall in the 24-Parganas occurred in September 1900, when, for four days, a shallow depression lay over South-West Bengal. During these four days the fall was 14 inches at Sāgar Island, 32 inches at Diamond Harbour, 30 inches at Calcutta and 25 inches at Barrackpore. Scarcely less remarkable was the precipitation in the first week of June 1913, when the rainfall was 17·38 inches at Alipore, 19·53 at Basīrhāt, 11·53 at Barrackpore, 13·73 at Bārāset and 17·18 inches at Diamond Harbour. Heavy downpours also occur in short periods, chiefly accompanying thunderstorms. At Calcutta, slightly over an inch of rain has fallen within 10 minutes, equivalent to a rate of nearly 7 inches per hour, and in May 1913 there were three inches in 20 minutes during a thunderstorm.

The severe cyclones of the transition periods, *i.e.*, in May and, later, in October and November, occasionally enter the north-west angle of the Bay of Bengal, in which case they move northwards into South-West Bengal. Serious danger arises when a storm wave occurs in combination with a high tide, as happened in the historic cyclone of 5th October 1864.

CHAPTER II.

HISTORY.

References to this portion of the Gangetic delta in the *Mahā-bhārata*, the *Raghuvansa*, and some of the *Purānas* show that at the dawn of history it lay between the kingdom of the Suhmas, in Western Bengal, and that of the Vangas, in Eastern Bengal, the boundaries of which were ill-defined and varied according to the power of their kings. The Suhmas lived near the sea coast on a great river, with marshes full of canes, which has been identified with the Bhāgīrathi,* while the Vangas are described as having fleets of warboats and a strong force of elephants. At the time of the *Raghuvansa* the country appears to have been subject to the Vangas, for that work tells of the defeat of the naval forces of the Vangas by Raghu, who "established pillars of victory on the isles in the midst of the Ganges." These islands, in all probability, represented the present area of the 24-Parganas, which was still a fen-land intersected by rivers and full of morasses. Ptolemy's map of the second century, A. D., indeed, shows the south of the delta as cut up by rivers and estuaries to such an extent that it was practically a collection of islands. In the seventh century, A. D., the district may have formed part of the land of Samatata, which is described by the Chinese pilgrim Hiuen Tsiang (Yuan Chwang) as a low-lying country bordering on the sea and rich in crops, flowers and fruits. Its climate was soft, its inhabitants were small of stature, of black complexion, and hardy. Buddhism was represented by 30 monasteries and 2,000 priests, but the number of Hindu temples was far greater. General Cunningham was of opinion that Samatata included the whole of the delta between the Bhāgīrathi and the main stream of the Ganges, its capital being at Jessore;† but the distances given by Hiuen Tsiang seem to indicate that the site of this city must be identified either with Dacca or with some place in the Bikrampur *pargana* of the Dacca district. However this may be, Samatala appears to be merely another name applied to Vanga on account of its flat, low coast-line; and as it is described as being 3,000 *li*, or 500 miles, in

EARLY HISTORY.

* Monmohan Chakravarti, *Notes on the Geography of Old Bengal*, J. A. S. B., 1908, pp. 284-5.

† *Ancient Geography of India* (1871), pp. 501-3.

circuit, as low and moist, and as situated along the sea coast, there is nothing improbable in the hypothesis that it included part of the present district. Samatata is also mentioned in an inscription of Samudragupta (*circa* 360 A. D.) as a tributary frontier kingdom of the Gupta empire, and Yasovarman of Kanauj (*circa* 731 A. D) is credited with the conquest of Vanga in the *Prākrit* poem *Gando-vāho*.

Nothing definite, however, is known of the district until the end of the fifteenth century, when a few details may be gathered from a Bengali poem of Bipradāsa, dated 1495, and from the *Ain-i-Akbari*, which embodies the rent-roll drawn up by Todar Mal in 1582 : these, it may be mentioned, are the earliest works extant in which Calcutta is referred to. The poem of Bipradāsa, which describes the voyage of a merchant called Chānd Saudāgar from Burdwān to the sea, mentions several riverside villages from Bhātpāra to Bāruipur. It tells us that he passed by Ariādaha, on the east, and Ghusuri, on the west, and that he then kept along the eastern bank of the river and passed Calcutta. Thence Chānd proceeded down the Adi Ganga, which, as mentioned in the previous chapter, was formerly the channel by which the waters of the Ganges found their way to the sea. From the *Ain-i-Akbari*, we find that the district formed part of the *sarkār*, or revenue division, of Sātgāon, which extended from near Sāgar Island on the south to a little above Plassey on the north, and from the Kabadak river on the east to beyond the Hooghly on the west. The greater portion lay east of the Hooghly, within what are now the districts of the 24-Parganas and Nadia. Calcutta (Kalikāta) is mentioned, with two other *mahāls*, as paying (in combination with them) a revenue of a little over Rs. 23,000 ; and several of the present *parganas* are found in the list of *mahāls*. Four of these lie along the northern edge of the Sundarbans, viz., Hāthiāgarh, Mednimal (close to Canning), Maihāti and Dhuliāpur. From the fact that these *parganas* were assessed to revenue, Dr. Blochmann is of opinion that the list of *mahāls* given in the *Ain-i-Akbari* "enables us positively to assert that in 1582 the northern outskirt of the Sundarbans, so far as it lies within the Presidency division, corresponded almost exactly to the northern boundary of the jungle marked on modern survey maps."*

PRATAP-ADITYA. The actual ruler of the Sundarbans at this time appears to have been a Hindu chief called Pratāpāditya, one of the chiefs known collectively as Bārah Bhuiyās, who were nominally

---

* *Statistical Account of Bengal,* Vol. I, p. 381.

vassals of the Emperor, but really enjoyed independence in the south and south-east of the Gangetic delta. Akbar's armies were occupied in quelling a dangerous military revolt and in campaigns against the turbulent Afghāns, who were in a state of constant rebellion and had actually made themselves masters of Orissa and part of Bengal. He had consequently not been able to reduce to complete subjection this outlying part of his dominions, and the Bārah Bhuiyās, secure in the retreat which the great waterways and swamps secured them, were for a time able to defy his power. There were in Bengal, says Du Jarric,* prefects of twelve kingdoms, who obeyed no one, paid no tribute and displayed a royal splendour. They did not, however, call themselves kings, but Boiones—which is merely a Latin translation of Bhuiyās.

A halo of legend attaches to Pratāpāditya, who is regarded by Bengali Hindus as a national hero. Briefly, tradition relates that his father, Bikramāditya, established the fortunes of the house and made his capital at Iswarīpur, now a small village in the Khulna district about 12 miles south of Kāliganj; it was also known as Yasohara, of which the colloquial equivalent is Jasor. He was eventually supplanted by his son Pratāpāditya, who removed the capital to Dhumghāt, another place in the Sundarbans, the actual site of which is doubtful, but which cannot have been very far from Iswarīpur. Pratāpāditya extended the limits of his kingdom by conquest, till all the surrounding country acknowledged his rule. He declared himself independent of the Mughal Emperor, and such was his power and prowess that he defeated, one after another, the imperial generals sent against him. In course of time he became tyrannical, and assassinated his uncle, Rāja Basant Rai, and all but one of his children. Nemesis followed, for an army sent, or commanded, by Mān Singh, Governor of Bengal from 1589 to 1604, succeeded in penetrating the recesses of the Sundarbans. Pratāpāditya was taken by surprise, his capital was captured and he himself made prisoner. He escaped, however, the ignominy of being paraded a captive in the imperial city, for, preferring death to dishonour, he swallowed some poison that he kept concealed in a ring.

Pratāpāditya is identified with the king of Chandecan, or Ciandecan, who is mentioned several times in the letters of Jesuit missionaries who visited Bengal at the end of the sixteenth century. The first to come to Bengal were two priests named

* *Histoire des choses plus mémorables advenues aux Indes Orientales*, Bordeaux, 1608-14.

Fernandez and Josa, who, on their arrival at Hooghly in 1598, were invited by the king of Chandecan to pay him a visit. The account of their journey given by Fernandez shows that the route lay through the Sundarbans, the king holding his court at a place situated half way between Chittagong and Hooghly. The king's dominions, he says, were so extensive, that it would take 15 to 20 days to traverse them. The country had a great trade in bees' wax, which was produced in the jungles; it was infested by dacoits, and he and his companion encountered great dangers, both from them and from tigers, before they reached Chandecan. Next year Josa was joined at that place by another Jesuit priest named Fonseca, who wrote that the king received him with great kindness: indeed, no Christian prince could have behaved better. A church was even built at Chandecan, which was dedicated to Jesus Christ: this was the first Christian church erected in Bengal. A few years later, in or shortly after 1602, some Jesuit priests went to Chandecan with Carvalho, the Portuguese commander of the island of Sandwip. The king promised to befriend them, but meditated treachery. He summoned Carvalho to "Jasor," and there had him put to death,* while the Jesuit priests were driven out. It remains to add that Purchas also mentions the reception of the Jesuit mission by Pratāpāditya—"The king of Chandecan (which lieth at the mouth of the Ganges) caused a Jesuit to rehearse the Decalogue. . . . This king and the others of Bacola† and Arakan have admitted the Jesuit into their country."

Mr. Beveridge identifies Chandecan with Pratāpāditya's capital of Dhumghāt, which he places in the neighbourhood of the modern Kāliganj in the Khulna district.‡ Another theory is that it was Sāgar Island. The latter theory is accepted by Bābu Rādhākumud Mookerji, who in his *History of Indian Shipping*, points out that Chandecan was one of the chief centres of Hindu naval activity during Mān Singh's viceroyalty. "By far the most important seat of Hindu maritime power of the times in Bengal was that established at Chandikan, or Sāgar Island, by the constructive genius of Pratāpāditya, the redoubtable ruler of Jessore. Numbers of men-of-war were always to be found ready for battle

* The murder was intended to secure the favour of the king of Arākān. Kedār Ray, of Sripur (in the Dacca district), another of the Bārah Bhuiyās, had obtained possession of the island in 1602 and placed it in charge of Carvalho. An attempt was made by the king of Arakan to take it, but was unsuccessful.

† Bakla, which was ruled over by another of the Bārah Bhuiyās. It comprised portions of the present districts of Backergunge and Dacca.

‡ *Were the Sundarbans inhabited in ancient times?* J. A. S. B., Vol. XLV, 1876.

and in a seaworthy condition at that naval station. There were also three other places where Pratāp built his shipyards and dockyards: these were Dudhāli, Jāhājaghāta, and Chakrāsi, where his ships were built, repaired and kept.*" It is said that a Portuguese admiral of Pratāpāditya (named Rodda) defeated the Mughal forces in a battle fought at the confluence of the Adi Ganga and Bidyādhari; at this spot there is a group of temples, with an old image of Kāli, the foundation of which is ascribed in the *Bangādhipa Parājaya* (by Bābu Pratāp Chandra Ghosh) to Rāja Basant Rai, the uncle of Pratāpāditya.†

FIRST ENGLISH SETTLEMENT.

In the latter half of the sixteenth century the emporium for the sea-borne trade of Bengal was Sātgāon on the Saraswati, not far from the town of Hooghly. The Portuguese shared in this trade, but, owing to the shallowness of the upper reaches of the river, their heavily laden sea-going vessels could not go up there, but had to anchor off Garden Reach to load and discharge, their cargoes being taken up and down the Hooghly in country boats. On the arrival of the Portuguese fleet every year, a bazar of mat huts sprang up at Bator, on the east bank, to be burnt down on its departure and re-erected next year on its return. In consequence of the silting up of the Saraswati, Sātgāon was gradually superseded as a commercial centre by Hooghly, and some of its inhabitants migrated to the western bank. Four families of Bysākhs, and one of Seths, founded the village of Gobindpur (on the site of the present Fort William in Calcutta), and another market was established at Sutānuti, the site of which is now occupied by the north of Calcutta. The name of the latter place, which means cotton market, shows that it was a bazar for the sale of country-made yarns and cloth. In addition to the Bengali population, the new settlement appears to have attracted Armenian merchants, for a tombstone in the churchyard of St. Nazareth, an Armenian church in Calcutta, bears an epitaph to the memory of the wife of "the charitable Sookeas," who died in 1630. It was here that the English, under Job Charnock, sought a refuge after their abandonment of the Hooghly factory in 1686.

There had for some time been friction between the English and the Mughals, owing to the exactions of the latter and their interference with the Company's trade; and Hedges, the Company's Agent in Bengal, repeatedly urged that they should definitely break with the local government, and build a fort on Sāgar

* *History of Indian Shipping and Maritime Activity* (1912), p. 218.

† *Bengal Past and Present* (April 1908), Vol. II, p. 259.

Island for their own protection and the maintenance of their trade. In 1683 the Directors expressed themselves against the view of "our late Agent and some of our Captains, that there is no way to mend our condition but by seizing and fortifying one of those pleasant islands in the Ganges about the Braces." Not long afterwards other counsels prevailed, and the Directors recorded their opinion that the Mughal Governors, having "got the knack of trampling upon us, and extorting what they please of our estate from us, by the besieging of our factories and stopping of our boats upon the Ganges, will never forbear doing so till we have made them as sensible of our power as we have of our truth and justice." Eventually, they despatched a fleet of six ships and three frigates, under Nicholson, with orders to take on board all their officers in Bengal, send an ultimatum to the Nawāb, and, if he failed to give a satisfactory answer, seize Chittagong, which was to be fortified and placed under Charnock as Governor.

The arrival of the *Rochester* and one frigate at Hooghly, and also of some reinforcements from Madras, brought up the strength of the forces under Charnock to 400 men. Hostilities broke out on the 28th October 1686, and Charnock soon realized that with such a small command, and at such a distance from the sea, he could not hold out against the overwhelming army which the Mughals could bring up. Accordingly, after some fruitless negotiations, he embarked his forces* on the 20th December 1686, and, dropping down the Hooghly, sought refuge at Sutānuti. "During January, 1687, he erected some hovels on the river bank of Sutānuti, even hoped for permission to build a factory, and got the length of signing twelve articles with the Viceroy's agent, which confirmed the previous grants of trade to the English, customs free. But in February, the swamps having shrunk to their cold weather dimensions, the Viceroy put an end to parleys by sending an army to crush the new settlement. 'The country all up in arms around us, and without any hope of peace,' wrote Charnock, the English had again to take to their ships, and seek refuge seventy miles further down the river, where, amid the tidal flats and creeks of Hijili, its waters merge into the sea."† On the way they stormed and took the fort of Tanna (which stood on the site of the present Botanical Gardens), an exploit laconically described by Charnock in the words—"On the 11th February assaulted and took his fort at Tanna, with the loss only of a man's

* According to the *Riyazu-s-Salātin*, Charnock signalized his departure by burning part of the town by means of a lens, which he turned on it from the deck of his ship!

† Hunter's *History of British India*, Vol. II, p. 257.

leg, and some wounded." Further down the river they plundered and destroyed the Nawāb's granaries and salt depôts, and also captured and carried off all the vessels they met. At Hijili, Charnock held out against an army of 12,000 men until June 1687, when he secured a treaty by which the English were allowed to settle along the Hooghly, but were forbidden to go beyond the Tanna forts and had to restore all the vessels they had taken.

From Hijili, the English moved to Uluberia in the Howrah district, which was thought to be a good site for a settlement. "Your town of Ulluberreah," wrote the Court of Directors, "hath, we understand, depth of water sufficient to make docks and conveniences for the repairing of any of our biggest ships and is a healthful place. We hope you may so manage that place or town of Ulluberreah, which you have articled for, that it may in time become a famous and well governed English colony."* Eventually, however, it was decided to concentrate at Sutānuti, as appears from a letter, dated 30th September 1689, in which the Bengal Council referred to "our reasons for the altering of our opinion about Ullaberreah and pitching on Chutanuttee as the best and fittest on the main," and stated that they were satisfied that Uluberia had been misrepresented by those sent to survey it.

Accordingly, in September 1687, Charnock again came to Sutānuti and set to work to make a permanent settlement there. "He again opened 'negociations' for leave to build a factory, and meanwhile hutted the remnant of his troops on the high eastern bank. For a year he laboured at the double task of buying a permit from the Viceroy, and erecting a factory in anticipation of it. With infinite labour and endurance of misery, through the hot weather and drenching rains of 1688, he threw up a rough shelter for his ague-stricken followers, and began some poor defensive works. To him arrived on September 20, 1688, Captain Heath, with another reproachful despatch from the Directors, and orders to put the whole survivors on board ship and to sail for the conquest of Chittagong."† The expedition sailed in November 1688, but was a lamentable failure. Nothing was done at Chittagong, and the fleet sailed back to Madras, where "Charnock ate out his heart for fifteen weary months." In February 1690, the Emperor granted the English a new license for trade, and the Viceroy of Bengal issued a permit authorizing them to return to Bengal, while, in return for the payment of Rs. 3,000, Charnock received a guarantee that

* Hedges Diary, Vol. II, p. 75.

† Hunter's *History of British India,* Vol. II, p. 260.

their trade should be free and exempt from local exactions. "Having received His Highness' guarantee, Charnock and his refugees at Madras made their way through the monsoon tempests of 1690 to the Hooghly river. At length on Sunday, August 24, 1690, at noon, the weather-beaten band anchored, for the third time, in the long pool of Calcutta. With a poor guard of 30 soldiers all told, they scrambled up the steep mud bank, which was thenceforward without a break to grow into the British capital of India. They 'found the place in a deplorable condition, nothing being left for our present accommodation, and the rain falling day and night.' Charnock's own fellow-servants, huddled together on the malarious river bank, almost mutinied for a return to their houses and gardens in Hooghly town.* But the old man knew that the Company's goods could never be safe so far beyond the guns of its sea-going ships. He had had enough of fenceless factories, and he resolved to create for his masters a stronghold which should be a surer guarantee than any *farmān*, even if he perished in the attempt. He perished: but not until by two more years' of endurance he had founded Calcutta.

"They were two miserable years. The buildings which he set up with so much labour and peril in 1688 had been burned. Three ruined earth hovels alone remained on the high river bank, and the wretched band had to live in boats during the most unhealthy months of the year. Throughout the pitiless monsoon months of 1690, Charnock struggled on, erecting such shelter as he could 'with mud walls and thatched till we can get ground whereon to build a factory.' In the scorching summer of 1691, we still find him and his desponding followers dwelling in only 'tents, huts and boats'. It is no wonder that the weaker brethren continued to clamour for their 'profitable, easy old habitations' in Hooghly town. Nor it is surprising that Charnock sent home an incomplete cargo that year, for which the superior Council, amid the comfort and plenty of Madras, soundly rated him. Yet Calcutta grew. Its deep pool attracted the trade from the Dutch and French settlements higher up the river, and Armenians began to flock to a place where they felt safe. But the fever-haunted swamps which stretched behind the river bank exacted a terrible price for its prosperity. 'Death overshadowed every living soul.'† The name of Calcutta was

* The poverty of Charnock's resources may be realized from the fact that two men whom he sent to Hooghly had to supply him with such necessary articles as a pair of *gharas,* three large dishes and a dozen plates.

† Wilson's *Early Annals of the English in Bengal*, p. 208.

identified with Golgotha, the place of skulls. Within a decade after Charnock finally landed on the deserted river bank in 1690, it had become a busy mart with 1,200 English inhabitants, of whom 460 were buried between the months of August and January in one year.* The miseries of the fever-stricken band throughout 1690 and 1691 are not to be told in words."†

The mortality of the pioneer settlers is not to be wondered at when it is remembered that the only place on which they could build their houses was a narrow strip of land on the river bank, and that inland stretched a swampy jungle and brackish lagoons. It has before now been pointed out that the place seemed marked out by nature as unfit for human habitation. "It is stated by Marshman that the reason for Charnock leaving Uluberia was the unhealthiness of that locality; but he certainly did not gain much in that respect by the change. Though allowed by the Nawāb to choose any site below Hooghly, he selected perhaps the most unhealthy site on the whole river. The Salt Water Lake on the east left masses of dead, putrid fish as the water receded in the dry season, while a dense jungle ran up to where Government House now stands. The new settlement was situated about 160 miles from the sea. The south wind—the only mitigation of the fierce tropical heat which prevails from the end of March to October—blows over salt marshes and steaming rice lands on its way to the city. Its almost uniform dead level, with depressions lying below the level of high water in spring tides, renders it difficult to drain properly, while the soil on which the town is built possesses every quality which the site of human habitation ought not to possess."‡ The writer concludes that Charnock selected Sutānuti for "some inscrutable reason," while

* Hamilton's *East Indies*, II, 7, 8.

† Hunter's *History of British India*, II, 266-268.

‡ Sterndale's *Historical Account of the Calcutta Collectorate* (1885). Sterndale also points out that less than 200 years ago the entire site was described as "a place of mists, alligators and wild boars", and that even 80 years later "Chowringhee was out of town, and *pālki*-bearers charged double fare for going to it, while at night servants returned from it in parties, having left their good clothes behind them through fear of dacoits, which infested its outskirts. When we consider the labour and expenditure incurred in making modern Calcutta what it is, we must admit the appropriateness of the motto adopted by the Corporation for the city arms—*Per ardua stabilis esto*".

The Indian view of the salubrity of Calcutta was no more favourable. The author of the *Riyazu-s-Salātin* pithily summed it up as follows in 1788: "Its air is putrid, its water salt, its soil damp." Its climate for eight months in the year was very unhealthy, and the best that he could say of the remaining four was that they were "not very unhealthy."

It is almost superfluous to add that Calcutta has been purified and rendered sanitary, like other tropical cities such as Havana, Colon, Vera Cruz and Rio de Janeiro, which were once regarded as pest-houses.

popularly his choice is often ascribed to chance.* The story is that he was delighted with the charms of the place while smoking a hookah in the shade of a large *pipal* tree near the site of the present Sealdah station or of a *nim* tree near the present Nimtala Ghāt.

It has already been shown, however, that England had realized the insecurity of settlements planted in the midst of populous cities, and exposed to every outburst of hostility or caprice of the local Governors. The policy of securing a fortified post at or near the mouth of the Ganges had long been urged upon the Directors, and was at last accepted by them; and the passage from the Bengal Council letter quoted above makes it clear that the selection of Sutānuti was the result of deliberate judgement, it being held to be "the best and fittest place on the Main." Sutānuti was, in fact, chosen for commercial and strategic reasons. The Hooghly river tapped the trade of the Ganges Valley, and Sutānuti was situated at the highest point at which the river was navigable for sea-going vessels. It was, moreover, protected against attack by the river on the west and by morasses on the east, and it could be defended by the guns of the shipping.

OTHER EUROPEAN SETTLEMENTS.

The English were not the first European nation to settle in the district. The Portuguese are said to have occupied Tārdaha on the Bidyādhari, at the spot where Tolly's Nullah now joins that river, a century before the foundation of Calcutta. They combined piracy with trade, and Channel Creek, the branch of the Hooghly which separates Sāgar Island from the mainland, was known in the eighteenth century as Rogues River, from the Portuguese and Magh corsairs who infested it. The Dutch had established a factory for salting pork at Barnagore before the end of the seventeenth century, and, later, maintained a station at Falta for sea-going vessels. Streynsham Master, the President of Madras, who visited Bengal in 1676, states that the Dutch had a hog factory at Barnagore, where they killed about 3,000 hogs a year, and salted them for their shipping; while Hamilton, in 1706,

* Rudyard Kipling, for instance, writes in a *Tale of Two Cities* in *Departmental Ditties*:

"Thus the mid-day halt of Charnock—more's the pity—
Grew a City.
As the fungus sprouts chaotic from its bed,
So it spread—
Chance-directed, chance-erected, laid and built
On the silt—
Palace, byre, hovel—poverty and pride
Side by side;
And, above the packed and pestilential town,
Death looked down."

refers to the ladies of easy virtue who made their homes there. Tiefentaller also says that Barnagore was famous for its *bāfta* cloths, and Price, in his Observations, says the cloth factories there determined Charnock to choose Calcutta as the site for his new settlement.

The Dutch took little part in the political dissensions of the time, but occasionally had to assert themselves, in order to prevent their trade being stifled by the exactions of the Mughal officials, who frequently held up the flotillas of cargo boats proceeding up and down the river. In 1684, we find that a squadron of four Dutch ships from Batavia anchored at Barnagore, in order to bring the Mughal Governor to reason by means of a naval demonstration. It had the desired effect, for the embargo on their cargo boats was withdrawn. A little later they were again embroiled with the Mughal authorities, and withdrew from their settlements, but in 1686, when war broke out between the Nawāb and the English, they were again put in possession of their factory and bazar at Barnagore, and appear to have done a good trade. Admiral Stavorinus, who came on a visit to Bengal, from Batavia, at the end of 1769, states that only an under-officer of the Fiscal of Chinsura resided at Barnagore, but the Dutch flag was kept flying, and a house was maintained for the accommodation of any of their servants who happened to stay there. He writes—"The coarsest sort of blue handkerchieves are made here"; also—"Barnagore is famous on account of the great number of ladies of pleasure who reside there, and who pay a monthly recognition to the Fiscal of Chinsura for the exercise of their profession." Of the small settlement at Falta, where the English took refuge after the capture of Calcutta by Sirâj-ud-daula in 1757, he writes—"The Fiscal of Chinsura keeps one of his officers here to have an eye on the illicit and smuggling trade, that is, in such cases when matters have not been settled with the Fiscal, and a proper consideration made for his connivance."

In the early part of the eighteenth century, the ill-fated and short-lived Ostend Company appeared among the ranks of the competitors for the trade of Bengal. The merchants of Ostend, Antwerp and other towns in the Netherlands, had long been anxious to have a share in the commerce with the Indies, and in 1722 succeeded in obtaining from the Emperor of Austria (in whose dominions the Netherlands were then included) a charter authorizing them to form a Company, known as the Ostend Company, to trade with the East Indies. Shortly before they had secured this charter, one of the ships which they had sent out as a private venture arrived in the Hooghly, and, with the assistance of the French at Chandernagore, succeeded in getting a full cargo.

Before sailing for Europe, the captain applied to the Nawāb, Murshid Kuli Khān, for some land on which to erect a factory, in case a company was incorporated. His request was readily granted by the Nawāb, who, it is said, was not only anxious to increase the trade of the province, but also to introduce rivals to the English. He therefore assigned the Flemish the village of Bānkibazar, on the bank of the Hooghly. The name of this village has dissappeared from the maps, and its site can only be identified from old charts, which show that it was situated near Garulia and Palta, about 3 miles north of Barrackpore. Here the Flemish factors at first resided in houses made of mats and bamboos, but they afterwards erected brick dwellings, built a wall round their factory, with bastions at the angles, and excavated a channel leading to the river, with a depth sufficieut to admit sloops of a considerable burthen.

Misfortune dogged the company from the outset. The first ship sent by it to Bengal, the *Emperor Charles*, which carried 30 guns, was lost while going up the Hooghly. The greater part of the cargo was, however, saved, and the officers and crew succeeded in reaching Bānkibazar. Their success in trade* was regarded with great jealousy by the Dutch and English, who are said to have bribed the Faujdār, or Commandant, at Hooghly, to make a false representation to the Nawāb of the strength of the fortifications at Bānkibazar, and of the danger of allowing them to retain a place of such strength within a few miles of Hooghly. The Nawāb believed the report, and ordered the closure of the factory. The Flemish defied his order, and made ready for resistance. The Faujdār then sent a large force from Hooghly, under the command of Mīr Jafar, who invested the place from the land side and, expecting a long siege, threw up entrenchments to protect his men from the Flemish fire. The besieged, however, completely commanded the river, and only permitted such boats to pass as they pleased. The French at Chandernagore, moreover, while pretending to assist in negotiations for peace, smuggled in arms and ammunition. Among others, the Flemish captured the son of a rich Mughal merchant of Hooghly, whom they detained as a hostage. The other merchants of that town brought pressure to bear on the Faujdār to secure his release, and a truce was declared for a few days, at the end of which he was set free, on his father paying a heavy ransom. After this, the siege was vigorously pushed on, and supplies cut

* According to Stewart, they completely established the Ostend trade in Bengal by underselling the other Europeans. The *Riyazu-s-Salātin* picturesquely states that they set the cap of vanity on the head of pride, and bragged that they would sell velvet, wool and silk fabrics as cheap as gunny cloth.

off. The Indians all deserted and left the Europeans to their fate. The garrison was reduced to fourteen persons, but they offered such a desperate resistance, and served their guns with such accuracy, that none of the Mughal troops dare advance from their entrenchments. The Agent himself had his arm shot off by a cannon ball, and, realizing that further resistance was useless, embarked with the survivors at dead of night, and reached in safety one of the Flemish vessels lying in the river, in which he set sail for Europe. Next morning the Mughal troops took possession of the factory, and razed its fortifications to the ground.

In 1727, the opposition of the European maritime powers forced the Court of Vienna to suspend the Company's charter for 7 years, but, in spite of this, the factory at Bānkibazar appears to have been reoccupied, and private merchants occasionally sent out ships to India. Stewart tells us that, "as the Agent of the head of the factory in Bengal was a person of great activity and determination, he continued to furnish them with cargoes." In order to stop this illicit trade, the English despatched a squadron, with orders to enforce a blockade. The Commodore sent two of his ships to take two of the Ostend ships that were anchored between Calcutta and Bānkibazar. The Ostend vessels made no fight. The *St. Theresa*, the smaller of the two, struck her colours without firing a shot, and was taken to Calcutta; the other, slipping her cable, took refuge under the guns of the Bānkibazar factory, where the English did not venture to follow her, and escaped. In spite of all their misfortunes, the Flemish appear to have kept up the factory at Bānkibazar until 1744, when they were again expelled.* The Company itself after a precarious existence, prolonged by the desire of the Austrian Government to participate in the East India trade, became bankrupt in 1784.†

---

* *Bengal Public Consultations*, October 14, 1744. M. Law also refers, in a letter, dated 1756, to "the affair of the Ostend Company" in 1744. Isolated factors were stationed at Bānkibazar until 1744 (*Imperial Gazetteer of India*, II 466).

† Stewart's *History of Bengal* (1813), pages 422–26, and *Riyazu-s-Salātin* (translation by Maulavi Abdus Salum, 1904), pages 276—78.

There is considerable confusion about the Ostend Company in works that refer to the settlement at Bānkibazar. The *Riyazu-s-Salātin* ascribes its establishment and defence to the Danes. Stewart, while stating that it was owned by the Ostend Company, calls it the German factory, and its defenders Germans. Mr. S. C. Hill in *Bengal in 1756-57*, though he refers in the body of his work to the Ostend Company, enters it in the index as "the Emden Company or Prussian settlement", and speaks of the defence of Bānkibazar by the Emdeners. There is similar confusion about the date of the capture of Bānkibazar by the Mughals. Stewart gives the date as 1733, and Orme as 1744, while Alexander Hamilton, in *A New Account of the East Indies*, gives it as 1723. The last date must be accepted, for Hamilton's book was published in 1727, and the whole affair is attributed by the *Riyazu-s-Salātin* to the time of Murshid Kuli Khān, who died in 1725.

THE EIGHTEENTH CENTURY.

To revert to the history of the English settlement, the rebellion of Subha Singh, which broke out in 1696, seemed at one time to threaten its very existence. The rebels overran the country as far as the western bank of the Hooghly, captured Hooghly itself, and, crossing the river, succeeded in capturing Murshidābād. One band even approached Sutānuti and set fire to the villages near it, but the zamīndārs of the neighbourhood attacked and drove them off with the loss of ninety men. The English were forced to have a ship at anchor in the Hooghly, in order to prevent the rebel bands from crossing the river, and when a body of the insurgents besieged the Tanna fort, they lent the commandant a vessel to serve as a guard-ship, with the help of which he repulsed the attack. In their fear for the safety of the settlement, they applied to the Nawāb for permission to fortify their settlement, and on being told that they might defend themselves, set to work to build a fort.

In 1698, they obtained from Prince Azimu-sh-shān, grandson of Aurangzeb, who had become Nawāb of Bengal, permission to purchase from their existing holders the right of renting the three villages of Calcutta, Gobindpur and Sutānuti. For this concession they paid the Prince Rs. 16,000, and thereby acquired a definite status as zamīndār of the three villages, for which they paid an annual revenue of Rs. 12,000. They followed this up by an embassy to the Emperor Farrukhsiyar's court at Delhi, to procure the recognition of their rights, and permission to purchase property on the bank of the Hooghly. The Emperor granted the permission sought for, but it was, to a great extent, rendered nugatory by the opposition of the Nawāb.

In 1742, the English merchants were again exposed to the alarms of war, for the Marāthas were raiding the country on the west of the Hooghly, and their cavalry might even sweep down upon Calcutta. The *Tigress* was moored off Perrin Point, to keep guard over the river, and in 1743 the Marātha Ditch was commenced. This ditch, or moat, took off from the Hooghly, and was intended to stretch in a semicircle from Sutānuti on the north, to Gobindpur on the south, a length of about seven miles: the line planned for it corresponded roughly with what is now the Chitpur Canal on the north, and thence first south, and then west, with the present Circular Road. Beginning at the Hooghly on the north, three miles were completed in six months, but the alarm then subsided, and the ditch was never completed: the portion of it running eastward from the Hooghly was eventually utilized in the construction of the Chitpur Canal. The danger of a Marātha raid was not, however, really over, for the *Consultations* of March 1748, refer to the Marāthas having captured "Tanner's

Fort," *i.e.*, the Mughal fort at Tanna, but fortunately for the English they did not attempt to cross the Hooghly.

Of the internal state of the district at this time there is unfortunately little record. It is known, however, that the people were exposed, not only to the sudden fury of cyclones, but also to constant raids by piratical freebooters and slave-dealers. Such a cyclone burst on 30th September 1737, when Calcutta "looked like a place that had been bombarded by an enemy." Great damage was done to the shipping, all the boats and small craft were destroyed, and "the storm laid the whole black town, throughout the Honourable Company's bounds, in so much that hardly 20 thatched houses were standing next day."* The raids of Portuguese and Magh pirates had long been the terror of the people in riverain tracts. Streynsham Master states in his diary of 30th November 1676, that the fort of Tanna had been built to check the incursions of pirates from Arakan, and that, ten or twelve years before, they had carried off the people of the riverside villages to their slave market at Pipli, "in consequence of which none durst live lower than this place."

The Sundarbans were infested by these corsairs, and a chain had to be run across the Hooghly between Calcutta and Sibpur to prevent them extending their raids up the river.† According to the *East India Chronicle* for 1758, the Maghs, in February 1717, carried off from the southern parts of Bengal no less than 1,800 persons—men, women and children. They were taken to Arakan, where the king chose the artisans, about one-fourth of the number, to be his slaves; the rest were sold into slavery, at prices varying from Rs. 20 to Rs. 70, and set to work on the land. Slavery was also common in Calcutta, as we may realize from the remarks of Sir William Jones, Chief Judge of the Supreme Court, in 1785—"Hardly a man or woman exists in a corner of this populous town who hath not at least one slave child, either purchased at a trifling price, or saved for a life that seldom fails of being miserable. Many of you, I presume, have seen large boats filled with such children coming down the

---

* C. R. Wilson, *A Short History of Old Fort William in Bengal.*

† From the *Riyazu-s-Salātin* we learn that the officer in charge of the fort at Makhwa, or Tanna, kept an iron chain (every link whereof was ten seers in weight) ready alongside the walls, to block the passage of pirate boats coming up the river. The author gives a travellers' yarn about Charnock forcing a passage on his way to Hijili in 1687. The Commandant had received orders not to allow the English ships to pass, and had accordingly laid the chain from one bank to another. Charnock, however, cut through the chain with an English sword and sailed down the river.

river for open sale at Calcutta. Nor can you be ignorant that most of them were stolen from their parents, or bought, perhaps for a measure of rice, in time of scarcity."

THE FALL OF CALCUTTA AND RETREAT TO FALTA.

On the 20th June 1756, Calcutta was taken by Sirāj-ud-daula, and the tragedy of the Black Hole followed. Drake, the Governor, pusillanimously deserted his charge, and fled to the ships, not without risk, for several muskets were fired at him by the enraged men whom he abandoned to their fate, but none of them hit him. Even on the ships, Drake and the survivors were not safe, for when they attempted to make their way down the river, they were forced back by the guns of Fort Tanna, and two of the smaller vessels were driven ashore. They accordingly returned to their anchorage off Gobindpur, and remained there till the 24th, when they were joined by three ships from Bombay that had run the gauntlet. Emboldened by this reinforcement, they set sail, and after losing another vessel, which ran ashore off Budge-Budge, reached Falta on the 26th June. Falta, we are told, was then "a place of some importance, having a town with a large bazar, and was moreover the station of all the Dutch shipping."* At Falta the English remained for six months, until an avenging force had been collected, and they felt themselves strong enough to take the field. In the beginning of August they received a reinforcement of 230 men, mostly Europeans, under Major Kilpatrick, which had been sent from Madras before the news of the fall of Calcutta. The Europeans from the out-stations also made their way to Falta, as well as those who had escaped from Calcutta during the attack, of whom there must have been nearly one hundred; while the fleet was materially strengthened by the arrival of vessels from England, Bombay and elsewhere, that were bound for Calcutta.

"But though they were thus early enabled to muster so considerable a force at Falta, they were wretchedly provided with arms, ammunition, stores, and even with clothing; nor were they

---

* "It strains the imagination," writes Sir William Hunter in *A River of Ruined Capitals,* "to conceive that this green solitary place was once the last foothold of the British power in Bengal. A consultation held by the fugitive Council on board the schooner *Phœnix* relates how their military member had written 'a complimentary letter to the Nawāb,' who had done their comrades to death, 'complaining a little of the hard usage of the English Honourable Company, assuring him of his good intentions notwithstanding what had happened, and begging him in the meanwhile, till things were cleared up, that he would treat him at least as a friend, and give orders that our people might be supplied with provisions in a full and friendly manner.' To such a depth of abasement had fallen the British power—that power to which in less than a year the field of Plassey, higher up the same river, was to give the mastery of Bengal."

much better off for provisions. Their numbers, too, soon began to decrease at a fearful rate, especially in Major Kilpatrick's detachment, in which the mortality was very great. Partly from the absence of sufficient shelter on shore, and partly from fear of surprise, they all slept on board the vessels, where, from want of room, they were compelled to occupy the open decks. This exposure during the rainy season, coupled with bad food and other privations, brought on a malignant fever, which infected all the ships, and ultimately carried off a majority of the party, leaving the remainder in a wretchedly reduced and pitiable condition." When at length, in December 1756, the relief expedition arrived from Madras, under the command of Clive and Admiral Watson, "sickness and death had been so fearfully busy amongst them, that, out of the party of 230 men brought round by Major Kilpatrick, one-half had sunk into the grave, and only about thirty of those who survived were fit for active duty.* The remnant of the Bengal military force, from being more accustomed to the climate, appears to have fared better.†"

Though his full force had not arrived, Clive determined on an immediate advance. His first objective was the fort at Budge-Budge, which Mānik Chānd, who had been appointed Governor of Calcutta (now renamed as Alinagar), had strongly fortified and garrisoned: it mounted no less than eighteen guns. The expedition left Falta on the 27th December, and next day anchored off Māyāpur. Before sunset on the 28th December, Clive started, with the Company's troops, on a march overland, his object being to get to the north of Budge-Budge and cut it off from Calcutta, while the fleet bombarded it from the river. It was a long and weary march, for, there being no draught cattle, the soldiers themselves had to drag the field pieces,‡ and they were misled by their guides, who took them inland through a swampy tract, intersected by watercourses, in which progress was necessarily slow. It was not till 8 A.M. next day that they reached their destination, a depression situated nearly ten miles north-east of the fort at Budge-Budge, and about a mile from the river bank. Exhausted by the night's march, they lay down to snatch a few hours' sleep, and, there being no suspicion that the enemy was anywhere near, no sentries or picquets were posted. Scarcely had

* Orme, Vol. 2, page 120. Ives, page 99, states, that to the best of his remembrance not above thirty remained alive, and not above ten fit for duty.

† Broom's *History of the Bengal Army*, pages 73-75.

‡ According to Clive, the field-pieces were, after all, of little use, "having neither tubes nor port-fires, and wrong carriages sent with them from Madras." Malcolm, Vol. I, page 154.

they been asleep an hour when Mānik Chānd came down upon them with a force of 1,500 horse and 2,000 foot. The troops, completely taken by surprise, ran to their arms and formed line as the enemy came up. Fortunately, the latter not expecting to find the English asleep, had merely opened fire, and had not come to close quarters. They succeeded in capturing the guns, but lacked heart to push home an attack. Clive rallied his men, and a bayonet charge dislodged the Mughal troops from their commanding position. The Bengal Volunteers recovered the guns, and their fire forced the Mughal infantry to fall back upon their cavalry, after which the British troops advanced in line. The enemy stood firm for some time, until a shot from one of the guns passed close to the head of Mānik Chānd, who was mounted on an elephant. Alarmed at his danger, he gave the signal for retreat, and the whole body moved rapidly off towards Calcutta.

In the meantime, the Admiral's ship, the *Kent*, having outsailed the rest of the fleet, anchored before the fort, and opening a heavy fire, soon silenced that of the enemy and made a breach in the ramparts; but, the troops having endured so much fatigue already, it was determined to defer the assault until the next morning. They accordingly passed the remainder of the day without any further hostilities, and in the evening were joined by a party of 250 sailors, under the command of Captain King.

" At night all was perfectly quiet. Not a sound was to be heard, save the measured tread or occasional challenge of the sentries posted round the camp—a precaution not neglected a second time—when suddenly the whole force was aroused by shouts and firing in the direction of the fort. Several of the sailors, excited by finding themselves once more on shore, with the prospect of active employment, and a liberal allowance of liquor that had been distributed that evening, strolled out to take a look at the fort. One of these, named Strahan, more adventurous than the rest, got up close to the walls unperceived, and, finding that the guns had made a practicable breach in one of the bastions, could not resist the inclination to ascend it. On reaching the top, he found a party of the garrison sitting together smoking. Being armed, he immediately fired a pistol amongst them, and waving his cutlass shouted out, " The place is mine ", giving three hearty cheers at the same time. The enemy, recovering from their surprise, and perceiving that he was alone, immediately attacked him. He defended himself with great intrepidity and skill, until at last his sword broke off close to the hilt, when he would have been overpowered, had not some of his comrades, who heard his shouts, opportunely arrived to his assistance. A sharp conflict now ensued. The troops rushed to the fort to join in the attack, and the garrison, of which a considerable portion had previously retreated, fled in all directions. The only casualty on this occasion was that of Captain Dugald Campbell* of the Bengal Service, who, whilst leading on a company of the sepoys, was shot by some of the sailors, who mistook them for a party of the enemy. Captain Eyre Coote, commanding the detachment of King's troops that had landed, took possession of the fort that night."†

The sailor to whom the capture of the fort was due, received an unpleasant surprise next morning. "Strahan, the hero of the previous night's adventure, was brought before Admiral Watson, who, however much he might admire the individual bravery displayed, considered it necessary to show his displeasure at the breach of discipline that had been committed. On being called upon for an explanation of his conduct, Strahan replied—'Why, to be sure, Sir, it was I that took the fort, but I hope there was no harm in it.' The Admiral, scarcely able to repress a smile at the simplicity of the answer, expatiated on the consequences that might have ensued from such irregular conduct and finally dismissed him with a severe rebuke and a threat of punishment. Strahan, somewhat surprised at this turn of affairs, no sooner found himself clear of the cabin than he exclaimed—'Well, if I am flogged for this 'ere action, I will never take another fort by myself as long as I live, by G-d.' "† It is needless to say that no punishment was inflicted.

On the 2nd January 1757, Calcutta was recovered, and at the end of the month the Nawāb advanced to retake it, with an army of more than 40,000 men, against whom Clive could put into the field only 1,350 Europeans and 800 sepoys. With this small force he attacked the Nawāb, who had taken up a position‡ between the Salt Lake and the Marātha Ditch. The action took place near what is now the Circular Road in one of the morning fogs so common in and round Calcutta in the cold weather; and the English, after carrying the enemy's camp, lost their way. The battle was not, therefore, as decisive as it would otherwise have been, but the reverse was sufficient for Sirāj-ud-daula, who become alarmed for his own safety and the communications of his army.

---

* Warren Hastings married the widow of this officer.

† Broome's *History of the Bengal Army*.

‡ The English occupied an encampment north of Calcutta. "While the olonel (Clive) was in search of a proper place for an encampment, a wild buffalo ıan at his guard, and although the sepoy it attacked discharged his musket ball into its body, and received it on his bayonet, yet the creature killed the man and made off." (Ives, p. 110). The presence of a wild buffalo so close to Calcutta shows how close the jungle was to the infant city. The Salt Water Lake occupied a much larger area than it does now, and came up to within a mile of Calcutta.

Accordingly, on the 9th February, he signed a treaty, by which he restored to the English the goods and villages that he had seized, promised compensation for what had been damaged or destroyed, recognized all their former privileges, and permitted them to establish a mint*, and erect fortifications.

CESSION OF THE 24-PARGANAS.

On the 20th December 1757, Mīr Jafar, the new Nawāb of Bengal, made an assignment to the East India Company of the zamīndāri or landholder's rights over a tract of country known as the Zamīndāri of Calcutta, or as the 24-Parganas Zamīndāri, from the number of *parganas* included in it.† This tract lay chiefly to the south of Calcutta, and comprised an area of 882 square miles. The Company received only the zamīndāri rights, *i.e.*, the right to collect rents from the cultivators, with the jurisdiction of a zamīndār over them, and was subject to the obligation of paying to the Nawāb the land revenue assessed on the land. The grant did not confer a full proprietary status, which was made over in 1759 to Clive by a *sanad*, or deed granting him the 24-Parganas as a *jāgīr*, or military fief, in return for the services which he had rendered, more particularly in aiding to suppress the rebellion of the Emperor's eldest son, who ascended the throne under the name of Shāh Alam. By this latter deed all the royalties, dues and rents collected by the Company, in its capacity as landholder, and paid by it into the treasury of the Muhammadan Government, were made over to Clive, who thus became, for all practical purposes, a superior landlord over his own masters, the Company. Clive's claims to the property, as the overlord of the Company, were contested by it in 1764; and in 1765, when he returned to Bengal, a new deed was issued confirming the unconditional grant to him for ten years, with reversion afterwards to the Company in perpetuity. The deed, having received the Emperor's sanction on the 12th August 1765, gave absolute validity to the original *jāgīr* grant in favour of Clive, but limited its term to ten years, after which the 24-Parganas were to be transferred to the Company as a perpetual property. The sum of Rs. 2,22,958, which was the amount of annual land revenue assessed upon them when they were made over to the Company in 1757, was paid to Clive from 1765 until his death in 1774, when the full proprietary rights reverted to the Company.

* The first coin struck in the English mint was issued on the 19th August 1757 and bore the name of the Emperor.

† The 24 Parganas were—(1) Akbarpur, (2) Amīrpur, (3) Azimābād, (4) Balia, (5) Barīdhāti, (6) Basandhāri, (7) Calcutta, (8) Dokhin Sāgar, (9) Garh, (10) Hathiāgarh, (11) Ikhtiārpur, (12) Kharijuri, (13) Khāspur, (14) Maidanmal (or Mednimall), (15) Māgura, (16) Mānpur, (17) Mayda, (18) Munrāgācha, (19) Paikān, (20) Pechakuli, (21) Satal, (22) Shāhnagar, (23) Shāhpur and (24) Uttar Pargana.

In 1759, war broke out with the Dutch, who, afraid of the rich Bengal trade being entirely monopolized by the English, had begun to intrigue with the Nawāb, who for his part was anxious to have a counterpoise to the power of the English. In October 1759, seven Dutch ships sailed up the Hooghly, having on board 700 European and 800 Malay troops. "To allow the Dutch troops to land, and form a junction with the garrison at Chinsura, was to admit the establishment of a rival and superior force in the province, which, coupled with the conduct of the Nawāb, was to submit to the certain ruin of the English influence and power in Bengal. To prevent this, which could only be done by force, was to commence hostilites with a nation with which the mother country was at peace*." Clive resolved on a bold course. He ordered up the only three Indiamen that there were in the river, with a smaller vessel called the *Leopard*, to protect Calcutta, and reinforced the garrisons in the forts on either side of the Hooghly. The Dutch sent a remonstrance, recapitulating their grievances and threatening vengeance, if the English hindered them from coming up the river or searched their vessels. Clive replied that there was no desire to injure the Dutch trade, or interfere with their privileges, but, under existing treaties with the Nawāb, it was impossible to allow their vessels or troops to pass. He therefore referred them to the Mughal authorities, offering his services as a mediator. His coolness enraged the Dutch, who began hostilities without further parley. They seized several trading vessels, captured the *Leopard*, and, landing at Falta and Raipur, attacked and burnt the English factories. On the other hand, Colonel Forde marched north from Calcutta, and, on the 20th November, seized the Dutch factory at Barnagore, after which he crossed the Hooghly so as to keep the garrison at Chinsura in check and intercept reinforcements.

THE DUTCH WAR.

The Dutch fleet came slowly and cautiously up the river, for they had no pilots, and on the 23rd landed the troops at Sānkrail, after which they dropped down the river to Melancholy (Mānikhāli) Point, below which the three English ships lay at anchor. Next day the English ships attacked them, in spite of their superior strength, for there were seven ships in the Dutch fleet, four of which mounted 36 guns each. After a fight lasting only two hours, the Dutch commodore struck his colours, and all his captains followed his example, except the second in command, who cut his way through, and, the English ships being too crippled to pursue him, escaped to Kulpi, where, however, he was captured by two other English

* Broome's *History of the Bengal Army*.

ships, who were hurrying up the river to join the meagre naval defences. Short as the action had been, it was fierce while it lasted. One of the English ships received no less than 90 shots in her hull, while her rigging was cut to pieces. Not one of her men was killed, however, though several were wounded, the sailors having been screened from the enemy's fire by bags of saltpetre—a risky, but successful, device. The land force of the Dutch fared no better, being signally defeated at Bedarrah, near Chinsura ; and the war ended with a treaty, by which the Dutch promised to send away all but 125 of their soldiers, to restore their fortifications to their former condition, and never to allow more than one ship at a time to come up the river beyond Kulpi, Falta or Māyāpur without the Nawāb's express sanction.

TILMAN HENCKELL.

The exigencies of space forbid any but a brief mention of the work of reclamation and development carried out in the Sundarbans towards the close of the eighteenth century by Tilman Henckell, who was Judge and Magistrate of Jessore in 1781, and had jurisdiction over the Sundarbans tract to the south. "His acquaintance," writes Sir James Westland in his *Report on the District of Jessore*, "with every subject affecting his district was most intimate ; and no wrong was too remote for his energy to grapple with, no advantage too distant for him to strive after. The idea of his administration was that it was the duty of Government to procure the peace and comfort of the mass of the inhabitants, though it might involve some harm in respect of the Company's commercial interests. These views were a little too advanced for his age, for there was then too great an inclination, on the part of Government officials, to look upon the natives as born only to be a means of profit to the Company. Mr. Henckell was never unmindful of his employers' mercantile interests, but he always set this before him as his duty—to guard the then almost helpless natives from the oppressions to which they were subjected by the commercial officers of the Company, as well as by their own zamīndārs."

In the Sundarbans, Henckell inagurated a system of reclamation, which, after many vicissitudes, has converted large areas of forest into fertile rice fields. His object was to introduce a body of peasant proprietors, holding directly under Government, and with this purpose he granted about 150 leases in 1785. At the same time, he established three stations in the heart of the Sundarbans, in order to assist in their development by providing markets for the sale of produce and the supply of boatmen plying along the waterways. One of these markets was situated at Henckellganj (now corrupted into Hingalganj), at the junction of the Jamuna and Kālindi in this district: the other two lie in Khulna. Measures were also taken for the protection of the boat routes, which were infested by dacoits, and for the abolition of the toll stations set up by zamīndārs, at which illicit tolls were levied, and traders were subjected to various exactions. Lastly, Henckell stepped in to protect the *mahindārs*, or salt boilers, from the oppression of the *malangis*, or middlemen, with whom the Agent of the Raimangal Salt Agency entered into contracts for the supply of salt. The *mahindārs* worked on a system of advances, which resulted in their servitude; the *malangis* not only had the power to drive them to work, but insisted on receiving Rs. 20 for every Rs. 4 advanced. So powerfully did his benevolence and his personality impress the people, that in his lifetime he received divine honours. Henckellganj was called after him, because it was believed that his name would be sufficient to keep off the tigers which infested the place and carried off the workmen while the land was being cleared. According to the *Gazette** of 24th April, 1788—"It is a fact that the conduct of Mr. H. in the Sunderbunds had been so exemplary and mild towards the poor Molungees, or salt manufacturers, that to express their gratitude they have made a representation of his figure or image, which they worship amongst themselves." Henckell, it may be added, died in 1800.

THE SEPOY MUTINY OF 1824.

The next most notable event in the history of the district was the mutiny which broke out among the sepoys stationed at Barrackpore during the Burmese War in 1824.

The sepoys had not enlisted to serve beyond the seas, but only in countries to which they could march. The regiments were, therefore, marched to the frontier station of Chittagong, and there assembled for the landward invasion of Burma. Several corps had already marched, and the 47th Bengal Infantry had been warned for foreign service, and was waiting at Barrackpore whilst preparations were being made for its march. Meanwhile, the British troops had sustained a disaster at Rāmu, a frontier station between Chittagong and Arakan, and the news, grossly exaggerated, reached Lower Bengal. Strange stories found their way into circulation as to the difficulties of the country to be traversed, and the prowess of the enemy to be encountered. The willingness which the sepoys had shown to take part in the operations beyond the frontier began to subside, and they were eager to find a pretext for refusing to march on such hazardous service. This

---

* The *Calcutta Gazette* at this time was not an official publication but a newspaper mainly devoted to European news and advertisements.

excuse was soon found. There was a scarcity of available carriage-cattle for the movement of the troops. Neither bullocks nor drivers were to be hired, and extravagant prices were demanded for wretched cattle, not equal to a day's journey. The utmost efforts of the commissariat failed to obtain the needful supply. In this conjuncture, a lie was circulated through the sepoy lines at Barrackpore that, as the Bengal regiments could not be marched to Chittagong for want of cattle, they, in defiance of their caste feelings, would be put on board ship and carried to Rangoon, across the Bay of Bengal. Discontent developed into oaths of resistance, and the regiments warned for service in Burma vowed they would not cross the sea. The 47th Regiment, commanded by Colonel Cartwright, was the foremost in the movement. That officer endeavoured, by conciliatory measures, to remove the cause of complaint; and Government offered to advance money for the purchase of such cattle as could be obtained.

These measures were without avail, and the regiment broke out into mutiny on parade on the 30th October. The sepoys declared that they would not proceed to Burma by sea, and that they would not march, unless they were allowed 'double batta.' Another parade was held on the 1st November, when the behaviour of the sepoys was still more violent. The Commander-in-Chief, Sir Edward Paget, a stern disciplinarian, next appeared on the scene. He proceeded to Barrackpore, with two European regiments, a battery of European artillery, and a troop of the Governor-General's Body-guard. Next morning, the rebellious regiment was drawn up in face of the European troops, but they still clung to their resolution. After some ineffectual attempts at explanation and concilation, the men were told that they must consent to march or ground their arms. Not seeing the danger, — for they were not told that the artillery guns were loaded with grape, and the gunners ready to fire—they refused to obey the word, and the guns opened upon them. The mutineers made no attempt at resistance, but broke at once, and, throwing away their arms and accoutrements, made for the river. Some were shot down; som were drowned. Many of the leading mutineers were hanged, and the regiment was struck out of the Army List.*

WAHABI RISING. Seven years later there was a rising of the Wahābis, or followers of Abdul Wahāb, an Arabian who appeared in the middle of the eighteenth century as a religious reformer, and founded a new Musalmān sect. The system which he set

* Kaye's *History of the Sepoy War*, Vol. I, pp. 266-269.

up was one of simple Puritanism, the object of which was to restore Islām to a purer form of faith, by stripping off the accretions which overlaid it. It claimed the right of private interpretation of the Korān, rejecting the authority of Hanifa, Malik, Shafi and Hanbal, the four Imāms, or founders of the orthodox schools which bear their name. The cult of the dead and the worship of saints were sternly interdicted, and last, but not least, the obligation to carry on *jihād*, or war against infidels, was proclaimed in no uncertain voice.

The apostle of the Wahābi faith in India was one, Saiyad Ahmad, a native of Rai Bareli, who was born in 1786, and became a convert during a pilgrimage to Mecca. On his return to India in 1823, inspired by the belief that he was the Imām of the 13th century of the Musalmān era, he began a crusade inveighing against the veneration of *pīrs*, or saints, denying the efficacy of offerings in the name of persons deceased, and preaching a holy war against infidels. In 1826, he announced that the time had come for a *jihād* against the Sikhs, and a fanatical war followed. The army and coffers of the Wahābis were replenished by supplies of men and money from Bihar and Bengal, and, in spite or reverses, the Wahābis overran the frontier, capturing Peshāwar in 1830.

The success of the Wahābis in the north emboldened the Wahābis of Bengal to rise. Their leader was Titu Miyān, a resident of the 24-Parganas, who in early life had been employed as a professional wrestler and *lāthiāl*. Having taken an active part in a riot he was imprisoned, and on his release went on a pilgrimage to Mecca, where he met, and became a disciple of Saiyad Ahmad. About 1827, he began secretly to preach the Wahābi doctrines in the neighbourhood of Bārāset. Fired by the successes of their brethren in the north, the Wahābis were ready to break out in open rebellion, when a match was laid to the train by a Hindu zamīndār, who imposed on each of his Wahābi tenants a tax of Rs. 2-8, which he described as a fine on beards. A riot ensued, in which a mosque was burnt down. This was followed by charges, counter-charges, fictitious suits, etc., and Titu Miyān realized that the psychological moment had come to proclaim the *jihād* to the enraged Wahābis.

"A series of agrarian outrages followed, ending in the insurgents entrenching themselves in a fortified camp, and defying and beating back the English authorities, with some slaughter. The whole of the country north and east of Calcutta, including the 24-Parganas, Nadia and Farīdpur, lay at the mercy of insurgent bands, between three and four thousand strong. The sectaries began by sacking a village in Farīdpur district, because one of the inhabitants refused to accept their divine mission. In Nadia district, a second village was plundered and a mosque burnt down. Meanwhile, contributions of money and rice were levied from the Faithful; and, on the 23rd October, the insurgents selected the village of Nārikelbāria, in the 24-Parganas, for their headquarters, and erected a strong bamboo stockade around it. On the 6th November they marched out to the number of five hundred fighting men, attacked a small town, and, after murdering the priest, slaughtered two cows, with whose blood they defiled a Hindu temple, and whose carcasses they scoffingly hung up before the idol. They then proclaimed the extinction of the English rule, and the re-establishment of the Muhammadan power. Incessant outrages followed, the general proceeding being to kill a cow in a Hindu village, and, if the people resisted, to murder or expel the inhabitants, plunder their houses, and burn them down. They were equally bitter, however, against any Muhammadan who would not join their sect; and, on one occasion, in sacking the house of a wealthy and obdurate Musalmān, varied the proceedings by forcibly marrying his daughter to the head of their band.

"After some ineffectual efforts by the district authorities, a detachment of the Calcutta Militia was sent out, on the 14th November, against the rebels. They, however, refused all parley, and the officer in command, being anxious to save bloodshed, ordered the sepoys to load with blank cartridge. The insurgents poured out upon us, received a harmless volley, and instantly cut our soldiers to pieces. All this took place within a few hours' ride from Calcutta. On the 17th, the magistrate got together some reinforcements, the Europeans being mounted on elephants. But the insurgents met them, drawn up in battle array, a thousand strong, and chased the party to their boats on the river, cutting down those who were slowest in retreat. It now became necessary to deal with the rebels by means of regular troops. A body of Native Infantry, with some Horse Artillery, and a detachment from the Body-guard, were hastened out from Calcutta. The insurgents, disdaining the safety of their stockade, met the troops upon the open plain, with the mangled remains of a European, who had been killed the previous day, suspended in front of their line. A stubborn engagement decided their fate. They were driven back pell-mell into their entrenchment, and the fortified camp was taken by storm. Titu Miyān, the leader, fell in the action. Of the survivors, three hundred and fifty in number, a hundred and forty were sentenced by the Court to various terms of imprisonment; and one of them, Titu's lieutenant, was condemned to death."*

The last historical event to be recorded is the outbreak of the Mutiny of 1857 at Barrackpore.† At that time Barrackpore was the head-quarters of the Presidency Division of the Army, which was under the command of General John Hearsey, an experienced officer, who had an intimate knowledge of the manners and customs of the sepoys and spoke their language with great fluency. It was garrisoned by four native regiments, viz., the 2nd Grenadiers, the 43rd Light Infantry and the 34th and 70th Native Infantry. As is well known, it had been decided to introduce the Enfield rifle in place of the musket with which the sepoys had hitherto been armed. Cartridges with greased paper were manufactured at the arsenal in Fort William for use with the rifle, and a depôt for instruction in handling the new weapon was started at Dum-Dum. It is clear that the sepoys under instruction soon suspected that the grease used in the paper was made of the fat of pigs or cows, or both, and that their officers learnt of their suspicions. One day a *khālāsi* of the Dum-Dum magazine asked one of the sepoys for a drink of water from his *lota*. The sepoy refused, saying that the vessel would be contaminated by the lips of a low caste man. The *khālāsi* retorted that the sepoy would soon be deprived of his caste, for the Government was busy manufacturing cartridges greased with the fat of cows or swine, which the sepoys had to bite before loading. On hearing of this, the officer in command of the musketry depôt at Dum-Dum paraded the men and asked if they had any complaints to make. Two-thirds of them stepped to the front, and, respectfully protesting against the mixture used for the cartridge paper, asked that wax and oil might be substituted. Reports of these two significant occurrences were submitted to General Hearsey, who on 24th January forwarded them on and recommended that the sepoys themselves might be permitted to make up the cartridges with ingredients obtained from the bazars. His suggestion was accepted, but in the mean time rumours that they were to be forced to become Christians had obtained credence among the sepoys and a mutinous spirit was abroad. Proof of their

MUTINY OF 1857.

---

* *The Indian Musalmāns*, by W. W. Hunter, pp. 45-47; 2nd Ed., 1871.

† This account is compiled from Forest's *History of the Indian Mutiny*, Kaye's *History of the Sepoy Mutiny* and the *Red Pamphlet*. The last was published in 1857 under the title "*The Mutiny of the Bengal Army*, by one who has served under Sir Charles Napier." The book, which is somewhat rare, gives a graphic contemporaneous account of the occurrences at Barrackpore.

unsettled state was afforded by incendiary fires at Barrackpore, in one of which the telegraph station was burnt down. On the night of 5th February there was a secret meeting of the men of all the regiments at which they declared that they were willing to die for their religion and discussed plans for plundering the station and killing all the Europeans.

General Hearsey, in reporting this, pointed out that the native officers were of no use. "In fact, they are afraid of their men and dare not act; all they do is to hold themselves aloof, and expect that by so doing they will escape censure as not actively implicated. This has always occurred on such occasions, and will continue to the end of our sovereignty in India." In order to hear what the men themselves had to say, he instituted a court of inquiry, at which evidence was taken of the objections to the new cartridges. He found that their suspicious were so deeply rooted, as to be ineradicable, and recommended, as the only possible way of allaying the unrest, that the use of the greased paper should be discontinued and that the cartridges should, if possible, be made up of the paper which had hitherto been used for the musket cartridges. He further paraded all the regiments on the 9th February and made a speech to them in which he explained the absurdity of the belief that Government intended to force them to become Christians or wished in any way to interfere with their caste or religion. His words seemed to have a good effect, but the men were again thrown into excitement by the news that on the night of 27th February, the 19th Native Infantry had mutinied at Berhampore. The sepoys of that regiment had not actually committed any act of violence. They retired to their lines when ordered to do so by the Colonel, and they fell in on parade next morning without any symptom of insubordination. It was decided to punish the regiment by disbandment, and there being only one European regiment between Calcutta and Dinapore, a steamer was sent to Rangoon to bring Her Majesty's 84th Regiment.

General Hearsey again addressed the regiments on parade, on the 17th March; but it was plain, as the month drew to a close, that the hopes of the speedy subsidence of the unrest would be disappointed. "For when the troops at Barrackpore knew that the 19th were to be disbanded, and that an English regiment had been brought to execute the punishment, they believed, more firmly than they had believed at the beginning of the month that other white regiments were coming, and that the Government would force them to use the abnoxious cartridges, or treat them like their comrades that were marching down from

Berhampore to be disgraced. So the great terror that was driving them into rebellion grew stronger and stronger, and as from mouth to mouth passed the significant words, '*Gorā-log āyā*'—'the Europeans have come,'—their excited imaginations beheld vessel after vessel pouring forth its legions of English fighting men, under a foregone design to force them all to apostatize at the point of the bayonet."*

On the 29th March, it was reported to Lieutenant Baugh, Adjutant of the 34th Regiment, that one of the men of his regiment, Mungul Pandy† by name, was marching up and down the lines, armed with a loaded musket, calling upon his comrades to rise, and declaring that he would shoot the first European he came across. Lieutenant Baugh mounted his horse, and, with a pair of loaded pistols in his holsters, rode down to the parade-ground. Immediately in front of the quarter-guard the station gun was posted, from which the morning and mid-day salutes were fired. Mungul Pandy, on hearing of Lieutenant Baugh's approach, concealed himself behind this gun, took a deliberate aim and fired. The ball wounded the horse in the flank, and brought him with his rider to the ground. Lieutenant Baugh, however, quickly disengaged himself, and, snatching up one of his pistols, advanced on Mungul Pandy, who, finding himself unable to load his musket a second time, had taken up a sword which he had with him. Lieutenant Baugh fired and missed. Before he could draw his sword, the sepoy was on him, and with one blow brought him to the ground.

The Sergeant-Major of the regiment dashed in to his rescue and attempted to seize Mungul Pandy, but was also wounded and struck down. A Muhammadan orderly, Sheikh Pithu by name, who had followed Baugh from his quarters, now rushed forward and, holding Mungul Pandy, gave the two men time to get up and escape. All this took place not thirty yards from the quarter-guard of the regiment consisting of 20 sepoys under a jemadar. So far from attempting to rescue their officers, the jemadar forbade the men to stir. The men of the regiment moreover, who turned out in front of the lines and watched the whole occurrence, showed their sympathies lay with Mungul Pandy, turning their backs on Baugh, when he passed them, wounded and bleeding, and reproached them for not assisting him.

---

* Kaye's *History of the Sepoy Mutiny*.

† In the Hunterian spelling, Mangal Pānde. Pānde is a common name of Hindustani Brāhmans.

At this juncture, while Mungul Pandy was striding up and down, calling on his comrades to rise and die for their religion, General Hearsey, who had heard the firing, galloped down to the parade ground accompanied by his two sons. He at once ordered the jemadar of the guard to follow him and seize the mutineer, but the jemadar demurred, saying—"He is loaded and will shoot us." Then, according to General Hearsey's own account—"I again, shaking my revolver and pointing it partly towards him, sharply repeated the order." The jemadar looked askance at me and replied—"The men of the guard are putting caps on the nipples." I said, in a commanding and peremptory voice, "Be quick and follow me," and rode out in front towards the mutineer. The guard followed, my aide-de-camp on horseback close to the jemadar, armed with his revolver; my other son also close to the native officer similarly armed, Major Ross in rear of myself. As we approached the mutineer, we quickened our pace. My son, Captain J. Hearsey, called to me, "Father, he is taking aim at you, look out sharp." I replied, "If I fall, John, rush upon him and put him to death." At the last moment, however, Mungul Pandy turned his weapon upon himself, pulling the trigger with his toe. He fell severely but not mortally wounded and was taken off to hospital. General Hearsey then reproached the sepoys for having refused to move hand or foot to seize the man, to which they sullenly replied that he was mad with *bhāng* and had a loaded musket.

"On the 30th March, the 19th Native Infantry arrived at Baraset, about eight miles distant from Barrackpore. It had by this time transpired that they were to march into the latter station for the purpose of being disbanded: still, the behaviour of the men was respectful; and, in order to avert their fancied doom, they had sent in a petition to the Governor-General, offering, in case they were pardoned, to proceed at once to China, or to serve anywhere on land or sea. In short, they showed a repentant spirit and were never less inclined to join in a conspiracy against the State. On arriving on the morning of the 30th at Baraset, they found a deputation from the 34th awaiting their arrival. It has since transpired that these men made them a proposal—the result of their deliberations of the previous night—which it was well for us that they did not accept. On that very morning Her Majesty's 84th from Chinsura, a wing of the 53rd Foot from Dum-Dum, a couple of European batteries from the same place, and the Governor-General's Body-guard (native) from Calcutta had arrived at Barrackpore, and had been ordered to appear on parade with the native regiments at five o'clock on the following morning. The proposal made by the 34th to the 19th was to the following effect: that they should. on that same evening, kill all their officers, march at night into Barrackpore, where the 2nd and 34th were prepared to join them, fire the bungalows, surprise and overwhelm the European force, secure the guns, and then march on to, and sack, Calcutta. Had the 19th been as excitable then as they had shown themselves on the 25th of February, these views might possibly have been entertained; but they were repentant and ashamed of their former excess. That they were not thoroughly loyal is proved by the fact that the tempters were not reported. They were suffered to return unbetrayed, but their scheme was at once and definitively rejected.

"On the following morning, the 19th Regiment marched into Barrackpore. An order by the Governor-General in Council, in which their crime was recapitulated, their fears for their religion pronounced absurd, and their disbandment directed was read out to them, in the presence of the assembled troops," In recognition of their penitence and good conduct on the march from Berhampore, the sentence was not accompanied with any marks of disgrace. They were not stripped of their uniforms, and were provided with money to convey them to their homes. They were given the pay due to them and marched away under escort, cheering General Hearsey and wishing him long life.

In the case of the 34th, however, such clemency was out of the question. Mungul Pandy and the jemadar of the guard were hanged by order of Court Martial; the jemadar, when on the scaffold, confessed his guilt, acknowledged the justice of his sentence and adjured his comrades to take warning by his fate. An inquiry into the conduct of the regiment was instituted and the Court found that while the Sikhs and Musulmāns were trustworthy, no reliance could be placed on the Hindus. Lord Canning ordered the disbandment of the companies stationed at Barrackpore, and this order was carried out on 6th May. There was no mitigation of punishment, as in the case of the 19th. When they had laid down their arms, the uniforms which they had disgraced were stripped from their backs, and they were marched out of cantonments under an escort of Europeans, the number of the regiment being erased from the Army List. One incident was significant. They were allowed to keep their Kilmarnock hats, as they had paid for them. Before crossing the river, many of them were seen to take off their caps, dash them on the ground and trample them underfoot, to show their detestation of the Company's service.

SUBSEQUENT HISTORY. The subsequent history of the district is one of peaceful, but uneventful, development and progress. Its industries have grown, its communications have been improved and extended, the population has increased rapidly, and cultivation has expanded, more specially to the south. There the jungle has been driven back, and agricultural colonies are spreading fast, though facilities for communication are still scanty; while the riverain tract in the north has become a centre of large organized industries. These aspects of its modern history will be dealt with in later chapters.

ADMINISTRATIVE CHANGES. An account of the way in which the the East India Company obtained possession of the 24-Parganas has already been given in this chapter. Regulation, II, III and IX of 1793 defined the jurisdiction of the civil, criminal, and revenue courts established in the 24-Parganas, but it was expressly ruled that the jurisdiction of these tribunals should not extend to the town of Calcutta. The arrangements of 1793 continued in force till 1800, when the Civil Courts of the 24-Parganas were abolished, and their jurisdiction was made over to the Judges of Hooghly and Nadia. At the same time, the jurisdiction of the Justices of the Peace for the town of Calcutta was extended to the suburbs and places within a radius of twenty miles, so that they possessed a jurisdiction nearly concurrent with that of the Magistrate of the 24-Parganas. The *Diwāni*, or Civil Court was, however, re-established in 1806. Regulation X of 1808 gave to the Magistrate of the 24-Parganas the duties and powers of a Superintendent of Police; and in 1811 the offices of Judge and Magistrate were united in the same person. Regulation XIV of 1814 divided the 24-Parganas into two distinct districts (*zilas*); one of which consisted of the suburbs of Calcutta, and the other of the rest of the district outside the suburban limits. This separation was made in consequence of the increasing population of the suburbs, and placed the suburban division, comprising the thanas of Chitpur, Māniktalā, Tāzerhāt, Nauhazāri and Sālkhia (in the Howrah district, in charge of a separate Magistrate and Judge. In 1832 Regulation VIII abolished the suburban district, and reunited the thanas just mentioned with the 24-Parganas.

The revenue and civil jurisdictions of the district did not coincide for many years, owing to the fact that, at the time of the decennial settlement of 1787, the landed property in this part of the country was chiefly held by the Rājas of Burdwān, Nadia, and Jessore. For convenience of collection, the revenues of their whole estates were made payable to the treasuries nearest to their principal places of residence; while, for police and general administrative purposes, it was deemed expedient to divide the territory without reference to the boundaries of estates or the treasuries into which their revenues were payable. Originally the district contained only 444 estates, but in 1816, when some estates were transferred from Burdwān, the number had risen to 564; these estates were retransferred to Burdwân in 1862.

For a considerable time the district was divided into two parts, the Alipore and Bārāset divisions, each of which formed a separate magisterial district. The former comprised the territory originally ceded to the Company, the latter consisted of the joint magistracy of Bārāset, which included several *parganas* that were transferred from Jessore and Nadia in 1834. The Bārāset Joint Magistracy was abolished in 1861, in which year the district was distributed into the following eight subdivisions—Diamond Harbour, Bāruipur, Alipore, Dum-Dum, Barrackpore, Bārāset, Basirhāt and Sātkhira. The boundaries as then laid down were modified in 1861 and 1863. All villages lying to the west of the Hooghly were excluded, the northern limits of the district were extended, and the land to the south, between the Ichāmati and Kabadak rivers, was added from the Jessore district. At the same time, the Kabadak river was declared the eastern boundary of the district, with the exception of a small area to the west of the Kabadak (now included in the Jhingergāchha thana of Jessore), which was left in the Jessore district.

Of the eight subdivisions above mentioned, the Sātkhira subdivision was detached from the 24-Parganas and made part of the Khulna district, on its formation in 1882; the Bāruipur subdivision was abolished in 1883, and the Dum-Dum and Barrackpore subdivisions and 1893 The Barrackpore subdivision was, however, reconstituted in 1904 from portions of the Sadar (Alipore) and Bārāset subdivisions.

From 1816 the administration of the Sundarbans was governed by Regulation IX of 1816, which provided for the appointment of a Commissioner in the Sundarbans, and vested him with the duties, powers and authority of a Collector of Land Revenue. This arrangement was discontinued in 1905, when it was realized that the time had come to co-ordinate the administration of this tract with the general administration of the district. The appointment of a special officer in 1816 had been necessary, because, the country being extensive, wild and inaccessible, the work of developing its resources was beyond the capacity of the Collectors of the adjoining districts. It was now felt that the necessity of having an officer with independent powers had disappeared, and that it was desirable that this tract should be administered entirely by the Collectors of the districts concerned.

This was not a very great change, for the administration had long been conducted by the District Officer, with the exception of making settlements and holding enquiries to see if the conditions of the settlement leases had been carried out. The District Officer already controlled excise, education, police, crime, *chaukidāri*, the opening out of communications and other branches of administration, and it was now decided that he should also exercise control over the important matters connected with settlements. Accordingly, in 1905 the Sundarbans Act (Bengal Act I of 1905) was passed, by which Regulation IX of 1816 was repealed, the office of the Commissioner in the Sundarbans was abolished, and his functions were transferred to the Collectors of the three districts (24-Parganas, Khulna and Backergunge) within which the Sundarbans are comprised.

CHAPTER III.

THE PEOPLE.

GROWTH OF POPULATION.

Since 1872, when the first census was taken, the population of the 24-Parganas has increased by 852,656, or 54 per cent., and now numbers 2,434,104; the figures of each census are shown in the margin. Though the population as a whole has grown steadily and uninterruptedly, there have been considerable local variations. In 1881, when there was a net increase of 6·9 per cent. for the whole district, there was a decline in the north and east owing to the prevalence of malaria. Burdwān fever appeared there in 1861, and, though it was said to have died out after three years, the tract continued to be very unhealthy, and the Barrackpore subdivision had a loss of 9 per cent. and the Naihāti thana of 10½ per cent. In 1891 the district showed a further increase of 11·9 per cent., but several of the northern and central thanas remained stationary or lost population, the worst being Hābra, where there was a decline of 5·4 per cent. In the next decade (1891-1901) another increase of 9·9 per cent. was registered, in spite of the fact that the central and northern thanas showed no improvement. On the other hand, the riparian population grew by 12 per cent. owing to the development of the industrial towns along the Hooghly, and the rate of growth was twice as fast in the southern thanas, where the progress of reclamation in the Sundarbans attracted numerous settlers.

Year of census.	Total population.	Increase per cent.
1872 ...	1,581,448	...
1881 ...	1,690,771	6·9
1891 ...	1,891,288	11·9
1901 ...	2,078,359	9·9
1911 ...	2,434,104	17·1

Conditions between 1901 and 1911 were in favour of a further growth of population. The public health was good, the births exceeding the deaths by 100,000. The outturn of the crops was well up to the average during the first four year of the decade. In 1905 they were short owing to heavy but unevenly distributed rainfall, while the rainfall next year was deficient and the outturn was again poor. Consequently, in 1907 there was distress, to meet which agricultural loans and other relief measures were necessary. On the other hand, there was a continued and increasing activity in manufacturing and industrial centres, which led to an addition of 50 registered factories (*i.e.*, factories employing 50 hands or more) and of 75,000 employés. Considerable progress was also made in the reclamation of the Sundarbans, where agricultural colonies are growing rapidly in spite of the absence of facilities of communication. The north of the district has no such drawbacks, for areas which were without railway communication are now served by the Bārāset-Basīrhāt Light Railway, which was opened to traffic in 1905 and extended to Hasanābād in 1909; another line from Beliāghāta Bridge to Pātipukur was opened in 1910. The suburban traffic between Calcutta and stations in this district has also developed rapidly: in 1910 the number of season tickets issued to and from Sealdah was 31,766. The extension of the Calcutta electric tramway to Alipore, Tollygunge and Behāla has assisted in the development of those places, while the Port Commissioners' steamer service has popularized the riverain municipalities in

THANA, ETC.	Population, 1911.	Variation per cent, 1901-11.	Density per square mile, 1911.
SADAR SUB-DIVISION—	***756,348***	+ ***12·67***	***650***
Bāruipur	100,309	+ 10·49	1,056
Mātla	92,637	+ 15·49	138
Jaynagar	123,866	+ 14·72	2,064
Bhāngar	87,470	+ 14·56	781
Behāla	56,482	+ 0·93	1,364
Tollygunge	66,487	+ 19·42	2,462
Sonārpur	47,355	+ 8·17	1,155
Vishnupur	91,768	+ 9·26	1,257
Budge-Budge	95,974	+ 16·54	1,959
DIAMOND HARBOUR SUBDIVISION—	***515,725***	+ ***11·93***	***402***
Magra Hāt	140,019	+ 7·36	1,296
Falta	57,301	+ 9·35	1,146
Diamond Harbour	78,535	+ 11·11	1,287
Kulpi	152,224	+ 12·68	1,553
Mathurāpur	87,646	+ 21·50	91
BARRACKPORE SUBDIVISION—	***292,524***	+ ***41·69***	***1,540***
Barnagore ...	43,910	+ 13·62	5,489
Barrackpore	72,259	+ 102·80	5,558
Noāpāra ...	25,236	+ 48·55	4,206
Khardaha	26,465	— 13·66	1,321
Dum-Dum	22,887	+ 13·98	538
Naihāti	101,767	+ 55·18	1,060
BARASET SUB-DIVISION—	***292,791***	+ ***10·84***	***1,065***
Hābra	74,320	+ 5·21	722
Deganga	62,232	+ 12·72	841
Bārāset	156,239	+ 12·96	1,594
BASIRHAT SUB-DIVISION—	***429,476***	+ ***15·39***	***223***
Bāduria	141,940	+ 7·34	904
Basīrhāt	98,720	+ 13·57	1,085
Hāroa	68,433	+ 11·18	1,267
Hasanābād	120,383	+ 31·61	74
SUBURBS OF CALCUTTA—	***147,240***	+ ***45·28***	***14,724***
Cossipur-Chitpur ...	48,178	+ 18·23	15,056
Māniktala	53,767	+ 66·01	15,814
Garden Reach	45,295	+ 60·56	13,322

the Barrackpore subdivision. The result is that a growing number of clerks employed in Calcutta offices live outside the city and are daily passengers on the trains, trams or ferry steamers.

The total increase of population since 1901 is 355,745 or 17 per cent., nearly half of which may be ascribed to the increased number of immigrants (as shown in the margin), most of whom are attracted by the good wages offered in the mills along the Hooghly or for agricultural labour in the interior. Their number has risen by 176,000 since 1901, and they now constitute one-sixth of the total population. On the other hand, there has been a loss of 20,000 by emigration, and more than half of the increment of population must be attributed to natural growth. The Barrackpore subdivision has a phenomenal increase, representing 42 per cent., which is nearly entirely due to the influx of mill-hands – the proportion of males to females in the whole subdivision is 5 to 3. In none of the other subdivisions, whether industrial or agricultural, is the rate of growth under 10 per cent. There is no sign of a drain of the population to Calcutta; on the contrary, the development of suburban railways and river steamer services points to the fact that an increasing proportion of the workers in Calcutta prefer to have their homes outside the city.

	1911.		1901.	
	Male.	Female.	Male.	Female.
Actual population	1,305,700	1,128,404	1,092,916	985,443
Immigrants	275,479	126,144	154,953	70,414
Emigrants	72,097	67,148	63,017	55,943
Natural population ...	1,102,318	1,069,408	1,000,980	970,972

A special inquiry made by the Bengal Drainage Committee in 1906-07, showed that the noticeably malarious thanas are Dum-Dum, Khardah, Barrackpore, Noāpāra, Naihāti, Degānga and Hābra, and that the least malarious areas are Bhāngar, Mātla, Diamond Harbour and Budge-Budge. In the healthy thanas the rate of increase has varied from 11 to 17 per cent.; in the unhealthy thanas the natural loss of population by death or lowered vitality is counterbalanced by immigration. Five of the seven unhealthy thanas lie along the Hooghly in the Barrackpore subdivision, where mill-towns cluster closely together, and the effect of malaria is obscured by the shifting of population to industrial centres. Hābra has an increase of only 5 per cent., a rate which is only a little below that in the adjoining thana of Bāduria. The two thanas last mentioned lie in the extreme north-east of the district, and have advanced at a relatively slow pace, compared with the thanas immediately to the south of them, viz., Bārāset, Deganga and Basīrhāt, which have all benefited by the opening of the light railway and have grown at a uniform rate of 13 per cent. There has been even more rapid progress in the Sundarbans thanas to the south and south-east, where cultivation is rapidly spreading. Hasanābād, which has also been opened up by the railway, has a gain of 32 per cent., and Mathurāpur of 21 per cent.

Proportionately, the greatest growth of population has taken place in the Suburbs of Calcutta, *i. e.*, the three towns of Cossipur-Chitpur, Māniktala and Garden Reach, where it amounts to 45·3 per cent. The most progressive of these towns is Māniktala, which has added 66 per cent. to its numbers. It is closely followed by Garden Reach with 60·6 per cent., while Cossipur-Chitpur is content with the more modest advance of 18·2 per cent. The increase in Māniktala and Cossipur-Chitpur is nearly entirely due to the greater influx of immigrants, the extent of which may be gauged by the marginal figures. In Garden Reach the addition of 17,084 persons is partly the result of an extension of the municipal boundary, the added area having a population of 6,444, or one-third of the net gain. It is also partly due to industrial activity; at the time of the census no less than 10,460 males, or nearly two-fifths of the total male population, were employed in the mills and dockyards situated within the town. As regards the increase in the number of immigrants residing in Garden Reach, it should be explained that altogether 21,986 persons, or nearly half the population, returned their birthplace as Calcutta, and that only 2,797 were recorded as born in the 24-Parganas, whereas the corresponding figures in 1901 were 1,865 and 14,270 respectively. It appears certain that at this census a large number of persons, who were born in Garden Reach, returned their birthplace as Calcutta thinking that the town formed part of Calcutta and not being aware that for administrative purposes it is included in the 24-Parganas.

TOWN.	INCREASE SINCE 1901.	
	Total	Immigrants.
Cossipur-Chitpur ...	7,428	5,916
Māniktala ...	21,380	19,296
Garden Reach ...	17,084	28,557
Total ...	45,892	53,709

DENSITY. In the district as a whole there are 502 persons to the square mile, but the average is reduced by the uninhabited forest area in the Sundarbans, a labyrinth of tidal rivers and swampy forests, which extends over 1,711 square miles or more than a third of of the district; if this area is excluded, the mean density is 777 per square mile.

Numerous towns, with busy jute and cotton mills, stretch along the whole length of the Hooghly from Garden Reach northwards, but away from its banks, the population is almost entirely rural and devoted to agriculture. Density in the different subdivisions varies accordingly, being as high as 1,540 in the Barrackpore subdivision, which is a narrow riparian strip crowded with municipal towns, factories and mills. In the Diamond Harbour subdivision it is less than a third of this, and in the Basīrhāt subdivision there are only 223 persons per square mile. Both these subdivisions, however, lie to the south and merge in the Sundarbans. The difference between conditions in the north and south is even more plainly seen in the thana returns; no less than 19 thanas have more than 1,000 persons per square mile, the density rising to over 5,000 in Barnagore (5,489) and Barrackpore (5,558), while in two (Mathurāpur and Hasanābād), which extend into the Sundarbans, there are less than 100 per square mile.

The population clusters most thickly in the suburbs of Calcutta, where the density is no less than 23 per acre. There is not much difference in this respect between the three towns, there being 25 persons per acre in Māniktala, 23 in Cossipur-Chitpur and 21 in Garden Reach. There is, however, considerable disparity between the different wards, as shown in the margin. In Cossipur-Chitpur the most populous wards (Nos. 1 and 2) lie along the Hooghly. In Māniktala density gradually increases from north to south. In Garden Reach it is highest in the circle next to Calcutta, and steadily falls the further one goes from the city, the minimum being reached in the circle furthest from Calcutta.

Town.	Ward or Circle.			
	1	2	3	4
Cossipur-Chitpur ...	39	29	12	23
Māniktala	20	24	26	...
Garden Reach	22	19	13	10

MIGRATION.

Since 1901 the foreign-born population in the 24-Parganas has increased by no less than 176,000, and now amounts to 402,000, or 16½ per cent. of the total population. The immigrants who outnumber the emigrants by 262,000, are drawn mainly from Bihār and Orissa and the United Provinces. The latter province contributes 84,000, and the former 145,000, of whom 97,000 come from Bihār (chiefly from the districts shown in the margin), 34,000 from Orissa and 14,000 from the Chota Nāgpur Plateau. Eastern Bengal and Assam can claim only

District.	Number.
Sāran ...	26,026
Shāhābād ...	21,012
Patna ...	13,651
Gaya ...	12,387
Monghyr ...	10,139
Muzaffarpur ...	9,344

9,000, while the Central Provinces accounts for 2,000 and Madras for 5,000: in Titāgarh alone there are over 3,000 mill-hands from Ganjām and Vizagapatam. The great majority of the immigrants are employed in industrial and manufacturing concerns, but the reclamation of the Sundarbans in the south attracts a number of cultivators and labourers from Midnapore and also from Chota Nāgpur. The emigrants from the 24-Parganas do not spread far beyond its limits, all but 17,000 being enumerated in adjoining districts.

PROPORTION OF THE SEXES. In the population as a whole males outnumber females by 177,000, the excess being due to the influx of immigrants, who find temporary employment in the mills, factories etc., and leave their families at home. In the district-born population there is actually a small excess of females, amounting to 28,000, but among those born outside the district, and enumerated in it, there are two males to every female. The disparity between the sexes is most pronounced in the mill towns, where the population is largely foreign-born. In places such as Bhātpāra, Cossipur-Chitpur, Gārulia and Titāgarh, the males outnumber the females by two to one, but in the non-manufacturing towns the sexes are equally represented, or the female element predominates.

TOWNS. The 24-Parganas is the most distinctively urban district in Bengal, 548,514 persons, or 22½ per cent. of its population, being inhabitants of towns. There are altogether 26 towns (including two cantonments) as shown in the margin,

Town.	Population.	Town.	Population.
Māniktala ...	53,767	Bāduria ...	13,680
Bhātpāra	50,414	Hālishahar ...	13,423
Cossipur-Chitpu	48,178	South Dum-Dum	12,874
Garden Reach ...	45,295	North Barrackpore	11,847
Titāgarh ..	45,171	Rājpur	11,607
South Suburbs ...	31,533	Gārulia	11,580
South Barrackpore	27,605	Pānihāti ...	11,118
Barrackpore Cantonment.	*11,485*	Jaynagar ...	9,245
		North Dum-Dum	8,865
Barānagar ...	25,895	*Dum-Dum Cantonment.*	*3,818*
Tollygunge ...	18,433		
Basirhāt ...	18,331	Bārāset	8,790
Naihāti	18,219	Bāruipur ...	6,375
Kāmārhāti ...	18,015	Tāki	5,202
Budge-Budge ...	17,982	Gobardānga ...	5,070

of which two have a population of over 50,000, six of 20,000 to 50,000, twelve of 10,000 to 20,000 and six of 5,000 to 10,000; the average population is 21,097.

These towns may be divided into four groups. (1) The first consists of five towns adjoining Calcutta, which are suburban in character, and from a structural point of view can scarcely be distinguished from it, viz., Cossipur-Chitpur, Māniktala,

Garden Reach, South Suburbs and Tollygunge; taken together these five towns have added 40 per cent. to their population since 1901. (2) The second class consists of eight industrial towns which, with the exception of Budge-Budge, stretch northwards from Calcutta along the bank of the Hooghly, viz., Barnagore, the adjoining town of Kāmārhāti, Naihāti, the two contiguous towns of Hālishahar and Bhātpāra, Titāgarh, Budge-Budge and Gārulia. The increase in these towns has also been very great, averaging no less than 67 per cent. (3) There are three other towns along the Hooghly, viz., South Barrackpore, North Barrackpore and Pānihāti, which, however, are not industrial centres: of these, only South Barrackpore has shown an advance since 1901, which is partly accounted for by the increase of population in the Barrackpore Cantonment. (4) The remaining ten towns are situated inland, and are mostly rural in character; altogether, they have an addition of 6 per cent., the most substantial increases being found in South Dum-Dum, Bāruipur and Basīrhāt. The growth of Bāruipur may, however, be partly accounted for by an addition to its area.

The average town population has increased by 38 per cent. since 1901, and no other district in Bengal has such a record of urban growth. The whole riparian strip along the Hooghly north of Garden Reach is, in fact, becoming urbanized: already, owing to their growing density of population, it has been found necessary to subdivide the South Suburbs, South Barrackpore and Naihāti municipalities twice since their creation, so that they now constitute nine municipalities. Exceptionally large increases were returned for the mill towns in 1911. The aggregate population of seven has risen by 87 per cent., and Titāgarh has trebled, while Bhātpāra has more than doubled its population. The latter town has, indeed, increased five-fold since 1881 and is now the fifth largest town in Bengal. These large increases are accounted for by the influx of factory labour, as illustrated in the marginal table. The character of the population has changed so greatly owing to this influx, that some mill towns are now practically foreign towns planted in the midst of Bengal. In Bhātpāra, for instance, four persons speak Hindi to each person speaking Bengali: in Titāgarh 75 per

Towns.	Increase of population, 1901—1911.		Increase of operatives, 1901—1910.	
	Actual.	Per cent.	Actual.	Per cent.
Bhātpāra ... ... / Naihāti ... ... / Hālishahar ...	36,763	81	30,835	215
Titāgarh	29,106	181	21,649	162

cent. speak Hindi, 8 per cent. Telugu, and 4 per cent. Oriya, while 11 per cent. only speak Bengali. They are no longer the peaceful retired villages described as follows by Mr. Beverley in the Bengal Census Report of 1872: "Even in the neighbourhood of Calcutta the so-called townships are mere collections of villages—villages closely studded and densely populated, it is true, but still with small pretensions to be designated towns. The villages are grouped together for municipal purposes, and are thus shown in the census tables as towns; but cattle graze, and rice is sown and reaped, in their very midst."

VILLAGES. The census village corresponds to the mauza or survey unit of area, except in the Sundarbans, where it corresponds to the "lot," or grant of land. As a rule, there is not much difference between the mauza and the residential village, *i.e.*, a continuous collection of houses bearing a common name, with its dependent hamlets, but this is not so in tracts which were uninhabited at the time of the revenue survey and in which villages have since sprung up. Altogether 77½ per cent. of the population reside in villages, of which the number is 3,385, their average population being 557. Of the rural population 29 per cent. live in villages with under 500 inhabitants, 50 per cent. in villages with 500 to 2,000, 15 per cent. in villages with 2,000 to 5,000, and 6 per cent. in villages with over 5,000 inhabitants.

The village generally consists of small groups of houses scattered through the rice and jute fields: large compact villages, where periodical markets are held, are usually found only on the banks of the rivers. The villagers live, more or less secluded, in detached homesteads, surrounded by a belt of fruit trees or bamboo thickets: the screen of trees and jungle secures that privacy which the Bengali likes for his domestic life. The oldest villages are almost invariably found on the banks of the rivers or in their neighbourhood, where there are ridges of comparatively high land and of considerable extent. The central basins between such ridges are swampy and unhealthy, but as the population increases and the village site becomes more crowded, the people build their houses further away from the river bank on mounds artificially raised in order to keep them above flood-level.

The following account* of their external experience is reproduced from the volume of Hunter's Statistical Account of Bengal dealing with this district:—

"The dense mass of vegetation in which all Bengalis delight to shroud themselves, and which encircles the rich land-holder's

* First published in the *Calcutta Review*.

palace as well as the peasant's hut, is everywhere more or less productive. It is composed of the materials for food or for building--the cocoanut, the bamboo, the jack tree, and the mango. There may be seen the slender stalks of the betel tree, and the towering stems of the cocoanut above them, their long arms waving in the breeze; on the other side, probably, a thick garden of plantains, that curious link between the vegetable and the timber; in the background, an underwood of wild cane, twining itself round everything of firmer bulk; and a little further on, an undistinguishable mass of thorn, creepers and underwood of every shade, length and denomination. The husbandman must have his fruit tree and his bamboo, which yield him a return for no expenditure of labour but that required for gathering and cutting, his protection for the womankind, and his shade against the fierce sun of April and May. If he attains these primary objects, he is content, no matter how much miasma may be exhaled from the decaying vegetation, how much disease may lurk in that fair but deceitful mass of green foliage, how many reptiles and venomous snakes may be concealed in the unwholesome shades which surround his paternal inheritance. The sun, and gaze of the passing neighbour, must alike be excluded. Grant him this, and he will endure, with stoical fortitude, the periodical fever, the steamy heat of the rains and the fœtid water which stagnates in the pools whence he has dug the materials for his homestead site (*bhitā*), and which never feels the influence of the breeze and the light."

Major Smyth (in his Revenue Survey Report, 1857) gives **HOUSES.** the following account of the houses of the people:—"Their habitations, with some exceptions among the richer classes, are built of mud; the poorer classes often use brushwood, plastered with mud, to avoid the labour and expense of raising a wall; they are thatched occasionally with grass, but chiefly of paddy straw, and congregated in a dense mass of jungle. These huts have no apertures or windows of any kind beyond the doorway, the only ventilation being through the small space left between the thatch and the top of the wall, which also serves the purpose of a chimney. No whitewash within; on the contrary, the blacker they become with the smoke, the more comfortable they are considered. Exteriorly, they are washed, by the females of the family, with a mixture of cowdung and mud, which, when dry, gives them a somewhat cleanly appearance."

Writing seventeen years later, Sir Willam Hunter quoted this passage from Major Smyth's report, and added:—"Only the wealthy classes live in brick houses; the shopkeepers and the husbandmen generally in mud huts. The building materials of a shopkeeper's house consist of bamboo, timber posts, and thatching grass or *golpātā* leaves. The cultivator's hut is even more primitive and consists merely of bamboo and thatching grass or *golpātā* leaves, with mud walls. The number of rooms or huts to each household varies according to the condition of the family. A shopkeeper with a mother, wife and three children would have a hut with two or three verandahs for the dwelling of himself, wife and children; and another hut, to serve both as a cook-house and as the dwelling of his mother. A verandah is set aside, or sometimes a separate hut is built, for the purpose of receiving visitors and friends. The dwelling of an ordinary peasant, with the same sized household, would consist of a hut to dwell in, another small one for cooking in, and a cowshed."

At the present day, the general standard of comfort has decidedly risen. Some of the richer merchants and zamīndārs have large country houses, of which many may be seen along the first eight miles of the Grand Trunk Road; they are usually two-storied, and situated in the centre of extensive gardens. Similar houses have been built, here and there, by rich zamīndārs on their country estates. But setting aside these houses, which are, of course, exceptional, there has been a considerable increase in the number of brick houses occupied by the well-to-do, such as traders, members of the official and professional classes, etc., as opposed to the really wealthy. To go a step lower, the small trader or fairly well-to-do ryot usually possesses a homestead, enclosed by a mud wall or bamboo fence, and containing several huts, which serve as a dwelling house, cookshed, cowshed, granary or *golā*; the word hut, it may be explained, does not necessarily imply poverty or squalor, and is simply used in contradistinction to a brick house. They are usually built on a mud plinth, from one to three feet in height, and have wooden doors, set in a wooden frame-work, and often one or more windows. The walls are generally either formed of mud, pure and simple, which, when six inches or more in thickness, sets into a fairly solid and weatherproof wall, or of a bamboo framework, plastered with mud to keep it wind and water-tight. The roof is supported by wooden posts, and is thatched either with *san* grass, or with *golpātā*, *i.e.* the leaves of the *hentāl* or wild date palm. Sometimes the roofs are tiled, and of late years the use of sheets of corrugated iron for roofing has greatly increased. The dwellings of the poorer classes show every degree of difference from a comfortable homestead of this type down to the miserable huts described by Major Smyth. The well-to-do ryot or trader usually has a small patch of land attached to his house, on which vegetables grow, and an orchard of plaintains, date, cocoanut and betelnut palms, mangoes, jack trees, etc.

In the suburbs of Calcutta, Cossipur-Chitpur, Māniktala, etc., large portions are completely urban, and insanitary overcrowded *bustis* may be found rivalling those of Calcutta itself. All along the banks of the Hooghly, during the last 30 years, great mills and factories have sprung up, and the presence of a large population, drawn to their vicinity by the attraction of high wages, has brought the question of overcrowding to notice in places where, but for their existence, it would not have been thought of. These mills employ from a few hundred up to eight thousand hands, and so far as the labour force is recruited locally, the existing villages would suffice for their accommodation. The local population, however, is quite inadequate to supply the quantity of labour required, and a large number of the employés are immigrants from Bihār and the United Provinces. Most factories, therefore, have "lines" for a part of their labour force, and such "lines" are usually well built and drained, with a filtered water supply laid on and distributed by pipes and stands, and with decent latrine arrangements.

RELIGIONS.

From the marginal table, showing the numerical strength of the different religions found in the district, it will be seen that Hindus predominate, representing 63 per cent. of the total population, while the Musalmāns account for 36 per cent. The latter are relatively most numerous in the Bārāset subdivision, where they out-number the Hindus, and in the Basīrhāt subdivision, where there are ten followers of Islām to every eleven Hindus. The Animists consist almost entirely of aboriginal emigrants from Chota Nāgpur, who are mostly employed in reclamation of land in the Sundarbans. They include 5,538 Orāons and 5,896 Mundās; in addition to these, 6,517 Orāons and 7,296 Mundās were returned as Hindus. The group entitled "Others" includes several minor religions, professed for the most part by immigrants from outside the district, viz., 391 Sikhs, 207 Buddhists, 97 Jains, 88 Brāhmos, 26 Pārsis, 16 Confucians and 8 Jews.

Hindus	1,525,886
Musalmāns	879,547
Christians	16,027
Animists	11,811
Others	833

Vaishnavism.

The Vaishnava sect has a number of adherents in the district, and Khardah in the Barrackpore subdivision is one of its centres. The following account of it is extracted from Risley's *Tribes and Castes of Bengal* (Vol. II, pp. 343-44).

"Tradition has it that when Chaitanya died, his most zealous disciples formed a society consisting of six Gosāins, eight Kabirājs, and sixty-four Mahants to organize the new teaching and to make proselytes throughout Bengal. In course of time there arose three great centres of the faith—Khardah, between Calcutta and Barrackpore, the home of the Gosāins, or religious mendicants who traced their descent from Nityānanda, a devoted disciple of Chaitanya; Sāntipur, on the river Bhāgīrathi in Nadia, where Adwaitānanda or Adwaita, another zealous disciple, had settled; and Saidābād, in Murshidābād, the residence of Bīrbhadra, the son of Nityānanda. The Gosāins or 'Gentoo Bishops,' as they were called by Mr. Holwell, have now become the hereditary leaders of the sect. Most of them are prosperous traders and money-lenders, enriched by the gifts of the laity and by the inheritance of all property left by Bairāgis. They marry the daughters of Srotriya and Bansaja Brāhmans, and give their daughters to Kulins, who, however, deem it a dishonour to marry one of their girls to a Gosāin. As a rule, they are tall and well-made men, of light complexion, fair specimens of the Aryan type as found in Bengal. The Adwaitānanda Gosāins admit to the Vaishnava community only Brāhmans, Baidyas, and members of those castes from whose hands a Brāhman may take water. The Nityānanda, on the other hand, maintain that any such limitation is opposed to the teaching of Chaitanya, and open the door of fellowship to all sorts and conditions of men, be they Brāhmans or Chandāls, high caste widows or common prostitutes. The Nityānanda are very popular among the lower castes, and hold a leading position among Vaishnavas. A *pānjhā*, or silver hand, is the badge of the family. The Gosāins themselves worship Durga, but their disciples do not follow them in this. They observe the birthday of Chaitanya on the 13th Phālgun, the Govardhan Pūja on the first day of the new moon in Kārtik, and the Diwāli on the night of the Kālī Pūja. The Adwaitānanda Gosāins are highly esteemed by the upper classes of Bengal, and it is very unusual for a Brāhman or Baidya to enrol himself in the ranks of the other branch. They are said to be more sincere and more open to religious motives than the Nityānanda, and they avoid much scandal by refusing to initiate women.

"For the purpose of making proselytes and governing the Vaishnava church, Bengal is divided into circles, each circle having its own Gosāin, with whose jurisdiction no other Gosāin is supposed to interfere. Under the Gosāin is the *adhikāri*, or Superintendent, who acts as deputy, initiates disciples within a certain area, and collects fees. Under him again is the *faujdār*, called also *khuro*, or uncle (the Gosāin being the father), whose business it is to beat up proselytes, and whose activity is stimulated by a percentage of the fees. Lastly, comes the *chharidār*, or usher of the rod, who is merely the messenger of the *faujdār*. Persons who join the Vaishnava communion pay a fee of twenty annas, sixteen of which go to the Gosāin and four to the *faujdār*."

The Baishnab caste. The name Vaishnava, or, as it appears in the census returns, Baishnab, is also the designation of a distinct group or caste, which, however, differs from the ordinary caste in that it is not exclusive, but receives fresh accessions from outside. In this sense the name has a restricted meaning and does not connote a member of a religious sect only. Admission to the caste is easily gained, for the aspirant merely engages the services of a Gosāin, to whom he pays a fee, and with his help arranges to give the usual feast (*mahotsab*) to other Baishnabs. He eats with them, and is then a Baishnab. A large number of prostitutes are Baishnabs. It is frequently the case, however, that a woman of this class does not become a Baishnab until the near approach of death, or at least until she finds herself seriously ill. She sends for the Gosāin, pays her fee, and arranges with him about the *mahotsab*. She is then easy in mind as to her decent burial after death.

The caste, as a rule, receives recruits only from the lower orders, and members of the higher castes do not join their ranks unless they have been, or are in danger of being, expelled from their own caste. Unlike the main body of Hindus, they bury their dead, do not observe periods of mourning, and do not acknowledge the supremacy of Brāhmans. The Gosāins, whom they reverence as their spiritual leaders, are, it is true, Brāhmans, but the respect and honour which they enjoy are apparently due to their descent and not to their status as Brāhmans. Their position is, in fact, somewhat peculiar, for, as Brāhmans, they do not eat food cooked by the ordinary Baishnab.

Some popular beliefs. Mention may be made here of some popular beliefs, such as the worship of godlings of disease, local saints, etc., which do not conform to either Hindu or Musalmān orthodoxy, and in which both Hindus and Musalmāns join. Musalmāns may be seen bowing before the shrine of Keshabeswar (see the subsequent paragraph on pilgrimages) and do not fail to make offerings to Sitala, the goddess of small-pox, to Manasa, the goddess of snakes, and to Dakhin Dwār, the god of tigers. The elasticity of Hinduism similarly permits its votaries to adore Satya Pīr, (whom they Hinduize under the name of Satya Nārāyan)

Mānik Pīr, the god of cows, and Ola Bibi, the deity presiding over cholera—all godlings or saints of their lower class Musalmān neighbours.

Jagat Gauri.

In addition to Manasa, there is a snake goddess called Jagat Gauri, who is said to be the sister of Manasa, and, like her, is credited with power over cobras and other snakes. She is represented as seated on a throne, with a child on her lap, and her shrine is at Nārikeldānga. There a Barna Brāhman officiates at her worship, except in the case of Doms and Hāris, who sacrifice pigs to her. From the fact that the Hindus do not object to their doing so, provided the animal is slaughtered behind the altar, and not in front of it, it may, as Mr. Gait points out, be surmised that the control of the shrine has only recently been usurped by the Brāhmans. A fair is held to honour of this goddess on the fifth day of the moon in the month of Jyaishta.*

Tree worship.

A curious form of survival of tree worship, which is still practised in the district, under the name of Dhelāi Chandi, was discovered a few years ago by Mahāmahopādhyāya Hara Prasād Sāstri, who gives the following account of it in the *Journal of the Asiatic Society of Bengal*, Part III, 1902:—

"About twelve years ago, while taking a stroll in the fields to the east of Naihāti in the district of 24-Parganas, I was struck by seeing people picking up clods of earth and throwing them at a date tree close to the road on the left. In the course of half an hour I noticed four or five persons doing that. Being curious to know why they did so, I asked one of them, and he told me that a *Chandi*, a female deity—a form of Durga, Siva's consort—resided in the tree, and is propitiated by offerings of those lumps of clay. I use the word "offering", but he used the word *naibedya*, that is, an offering of uncooked eatables; so the *Chandi* is supposed to eat the lumps of clay. Unlike the propitiation of other deities, who grant boons enjoyable only in the world to come, the propitiation of this deity is followed immediately by a great relief, and the relief is that children crying at home are at once pacified. I had then a child about a year old whose cries often vexed the whole family, so I took a clod and threw it at the date tree. On approaching the tree, I marked two things—that the lumps of earth had covered several square yards of the ground to a height of eight or ten feet all round the tree, and that the tree was never tapped, so that it appeared like a giant among the often-tapped, indented, moribund date trees. What the consequences of my offering to the date tree were, I do not

* Bengal Census Report of 1901, p. 195.

remember, but I told the thing to several of my friends; and one of them informed me of the existence of a similar tree about half a mile north of Naihāti on the road leading from the Gaurīpur Mills to Mājipāra. Curiosity led me to pay a visit to that tree also, and I found the same thing there too.

"Ten years later, when I resolved to write on the subject of this curious worship, I thought it proper to pay visits to my old friends again. The new *kutcha* road from Naihāti to Amdānga had been made, and the *āsthān*, or seat of the deity, had fallen to the right and a few yards away from it. I had no difficulty in recognising the mound of earth. The old tree at the centre of the mound was dead, and its dried stump only occupied the old position, but by its side another tree had grown up to the height of the old one, and was enjoying the offerings of the passers by. On asking a rustic, whose house was situated in the next village, I learnt that, instead of lumps of earth, sweets are often offered,—sweets such as *sandes* and *rātāsā*,—and that the propitiation of the deity is followed, not only by the pacifying of the crying child, but also by other boons such as the birth of a child, the obtainment of a situation, success in litigation, etc. I asked him if any *mantras* were used with the offerings, and was answered in the negative. I also asked him whether there was any priest of the deity, and received a similar answer. Then I asked him what becomes of the sweets that are offered, and he said, they are picked up by cow-herd boys. The old man gradually became communicative, and told me of many miracles displayed by the presiding deity of the tree. He said that a neighbour of his once ventured to tap the old date tree (and he pointed out to me the mark of the tapping on its dry stump), but the man who ventured to commit such a sacrilege died in the course of a month by vomiting blood. He also told me of a hooded serpent which often came to the tree and which is really the *Chandi*. Thus in the course of ten years I found there were great changes in this very simple worship. The offerings had improved, the sphere of usefulness of the deity had expanded, a myth had grown up, and it only remained for a priest to appear in order to raise the worship to the dignity of a cult. When I visited the other date tree, I found the same improvements there too.

"Since my attention was directed to this form of newly growing tree worship, I have been informed of several other date trees in the same neighbourhood enjoying the same consideration and worship. There are two near the Kānchrāpāra station, one to its north-east on the *khāl* which is an old bed of the Jamuna, at a place named Kāntāganj, and the other to the south-west of the station and to the west of the locomotive workshops, near the Shāh-dighi, an old tank with huge banian trees, said to have been excavated by Malik Sāhib about two hundred years ago when he founded the old mosque at Bāg. There is a third tree near Mājipāra on the road which runs from the Gaurīpur mills to that village. There are a fourth at Chāndīgarh on the Amdānga road, a fifth on the old road leading to Nārāyanpur (now very little used because of the construction of a *pucca* road from the Kānkināra station to that village), and a sixth at Mandalpāra".

Adoration of Pirs. The adoration of Pīrs or Muhammadan saints is common among the lower class Musalmāns, and is not confined to them, for it is shared in by the more ignorant and superstitious among the Hindus. The Pirs are credited with supernatural powers, and their tombs, or *dargāhs*, are places of pilgrimage to which people resort for the cure of disease or the exorcism of evil spirits, or to obtain the fulfilment of some cherished wish. Vows are registered before them and offerings made, usually consisting of sweetmeats, which become the perquisite of the *mujāwir* or custodian of the tomb.

Machandāli Saif. One local Pīr of some local repute is Machandāli Saif, whose tomb is near Gangā Sāgar. Legend relates that the saint suddenly disappeared one day when he was being shaved by a barber, and returned shortly afterwards dripping with perspiration. On being questioned, he explained that a ship had run aground and that, having been invoked by the crew, he had pulled it back into deep water. The barber received the story with derision, and as a punishment for his disbelief, he and his family immediately died.

Ghāzi Sāheb. The most famous of the Pīrs of the district is Ghāzi Sāheb, whose thaumaturgic powers are celebrated in the following legend. The country in the Maidanmal or Mednimall *pargana* (south of Tolly's Nullah and containing Bāruipur), was formerly a dense jungle, overrun with wild beasts. A *fakīr*, by name Mobrah Ghāzi, took up his residence at a place of this jungle called Bāsra; he overawed the wild beasts to such an extent, that he always rode about the jungle on a tiger. The zamīndār being unable one year to pay his revenue, the Emperor ordered him to be arrested and brought to Delhi; on which his mother sought the *fakīr's* assistance in getting her son released. The *fakīr* promised to help her, and caused the Emperor to dream as follows: Mobrah Ghāzi, surrounded by wild beasts, appeared to him, saying that *he* was the proprietor of the Maidanmal jungle, and that the revenue due by the zamīndār would be paid from his treasures buried in the jungle. So saying, he asked the Emperor to release the zamīndār, threatening him with every misfortune if he disobeyed. The Emperor awoke and had the dream written down, but paid no attention to it. The next morning he ascended his throne, but instead of his usual attendants and courtiers, he found himself surrounded by wild beasts. This brought the subject of the dream to his mind, and in great fear he at once ordered the release of the zamīndār and sent him back to Maidanmal with an escort, instructing him at the same time to ascertain the spot where Mobrah Ghāzi's treasures were hidden, to dig them up, and to remit his revenue to the treasury at Delhi. On reaching home, the zamīndār informed his mother of all that had happened, and especially of the instructions regarding the treasure. She went immediately to Mobrah Ghāzi, who at her request pointed out the place where the treasures were buried, and ordered her to dig them up and take them away. He then mysteriously vanished. The mother and son dug them up next day, sent the Emperor his revenue, and transferred the remainder to the zamīndāri coffers. In gratitude to Mobrah Ghāzi, he wished to erect a mosque in the jungle of Bāsra for his residence; but he was prevented in a dream, in which the *fakīr* appeared to him, saying that he preferred living in the jungles, receiving offerings from all who came to cut wood, and that he required neither mosque nor house of any kind. The zamīndār then ordered that every village should have an altar dedicated to Mobrah Ghāzi, the king of the forests and wild beasts; and warned his tenants that if they neglected to make offerings before proceeding into the jungles, they would certainly be devoured. These altars to Mobrah Ghāzi are common in villages in the vicinity of the the Sundarbans; and wood-cutters never go into the jungle without invoking Mobrah Ghāzi's protection.

Exorcism of wild animals. A number of *fakīrs*, who call themselves descendants of Mobrah Ghāzi, gain their livelihood by the offerings made by wood-cutters and boatmen in return for their services in protecting them from the attacks of tigers The custom is for the *fakīr* to go with the wood-cutters to the spot where they have to work, and clear a bit of jungle, on which he marks out a circle, repeating charms and incantations. Within the circle he builds seven small huts with stakes and leaves. Beginning on the right, the first hut is dedicated to Jagabandhu, the friend of the world, the second to Mahādeva, the destroyer, and the third to Manasa, the goddess of snakes. Next to it a small platform is erected in honour of Rūpāpāri, a spirit of the jungle, and beyond this is a hut divided into two compartments—one for Kāli, the other for her daughter Kālīmāya. Then there is another small platform, on which offerings are made to Orpari, a winged spirit of the jungle; after this is a hut with two compartments, one being for Kāmeswari and the other for Burhi Thākurāni, and then a tree, called Rakshya Chandi (another name for Kāli), the trunk of which is smeared with vermilion: no offerings are made to it. Then come two more huts, with two compartments in each and flags flying over them. The first hut is reserved for Ghāzi Sāheb and his brother Kālu, and the next is for his son Chawal Pīr and his nephew Rām Ghāzi. The last deity propitiated is Bāstu Devata (the earth), who has no hut or platform, but receives offerings placed on plantain leaves on the ground. The offerings to the different deities are simple enough, consisting of rice, plantains, cocoanuts, sugar, sweetmeats, etc; *chirāghs* or small earthen lamps are lit; pots of water covered with mango leaves, and decorated with an image of the deity in vermilion, are put out; and flags are hung over the huts.

When everything is ready the *fakīr* has a bath, and returns wearing a *dhoti* provided for him by the wood-cutters, with his hands, arms, and forehead smeared with vermilion. Then, with hands folded before his face, he goes on his knees, bows his head to the ground, and remains in this attitude for a few seconds before each of the deities in succession, offering up prayers to each of them. After finishing his prayers, the *fakīr* proceeds to ascertain whether a tiger is present in the locality or not, by spanning his arm from the elbow to one of his fingers. If the span fails to meet a finger exactly, it is a sign that a tiger is present and the *fakīr* has to drive it off by repeating an incantation.

The *fakīr* then repeats charms for the protection of the wood-cutters and himself. After this, in order to close the eyes of the tiger, he repeats an incantation, beginning—"Dust! dust! The finest dust be on thy eyes, O tiger and tigress." Special charms are repeated if a tiger is seen in the jungle prowling anywhere near the wood-cutters, or is believed to be in their vicinity, or if the growl of a tiger is heard anywhere near the place where wood-cutting is going on. "That the *fakīr* is thoroughly believed in by wood-cutters," writes Mr. Sunder, "there is no doubt, and it is equally certain that his charms and exorcisms give them courage to enter the forests and embolden them to work there, notwithstanding the variety of dangers by which they are surrounded. Without him they would be utterly helpless. That his exorcisms and incantation have little effect has been proved, for it often happens that the *fakīr* himself, instead of the wood-cutters, is carried off by the tiger." It is believed that crocodiles, as well as tigers, can be commanded by the *fakīr*, who can make them rise or sink at will and can shut their mouths and prevent them attacking human beings.

When a tiger carries off a *mānjhi* of a boat, the helm used by him is removed from the boat and planted with the blade upwards on the spot where the man was killed, and a piece of white cloth, with some rice tied in a corner of it, is attached to the helm. When a boatman is killed by a tiger, his oar is planted, blade upwards, on the place where he was attacked, and a white flag, with some rice tied in a corner of it, is fixed to the oar. If any person attempt to remove either the helm or the oar and fail to draw it out of the ground by a single pull, it is believed that he will be killed by a tiger; but nobody ever interferes with the simple memorials to the dead, which are seen on the banks of streams and in the jungles throughout the Sundarbans.*

Ghāzi Sāheb and his brother Kālu are venerated both by Muhammadans and Hindus, and whenever any person desires to enter any jungle, he first bends to the ground, with hands folded before his face, and says: "In the name of Ghāzi Sāheb". Having done this, he goes into the jungle, believing that Ghāzi Sāheb will keep him perfectly safe.

The most celebrated place of pilgrimage in the district is Kālīghāt, which is situated on the bank of the old bed of the Ganges, two miles south of the southern boundary of Calcutta. The place derives sanctity from the legend that when the corpse of Siva's wife, Sati, was cut in pieces by order of the gods, and chopped up by the disc (*sudarsan chakra*) of Vishnu, one of her fingers fell on this spot. The temple is supposed to have been built about three centuries ago by a member of the Sābarna Chaudhri family of Barisa, who allotted 194 acres of land for its maintenance. A man of the name of Chandībar was the first priest appointed to manage the affairs of the temple, and his descendants, who have taken the title of Hāldār, are the present proprietors of the building. They have given up their priestly avocation and have amassed wealth, not so much from the proceeds of the temple lands as from the daily offerings made by pilgrims to the shrine. The principal religious festival of the year is on the second day of the Durga-pūja, when the temple is visited by crowds of pilgrims.

PILGRIM-

Kālīghāt.

---

* *Exorcism of Wild Animals in the Sundarbans*: J.A.S.B., Part III, 1903.

Ganga Sāgar. The greatest bathing festival of the year is that known as Ganga Sāgar, which is held at the southern end of Sāgar Island at the period of Makara Sankrānti in January. It attracts an immense number of pilgrims, who believe that by bathing at this sacred spot they wash away their sins. The sanctity of the place is explained by the legend mentioned in the article on Sāgar Island in Chapter XV, and is due to the fact that Sāgar Island marks the place where the waters of the Ganges mingle with the sea.

Keshabeswar. The temple of Keshabeswar, a form of Siva, at Mandira Bazar, or Bānchāpra, is visited by pilgrims on Mondays and Fridays, but the largest gathering takes place on the day of the Sivarātri festival The votaries, of course, are mainly Hindus, but many a Musalmān may be seen visiting the shrine in the hope of being cured of disease

Naihāti. Naihāti is visited by pilgrims from Eastern Bengal and elsewhere who are unable to meet the expense of a pilgrimage to Kālīghāt or who, for other reasons, prefer to make their way to Naihāti to have a bath in the Bhāgīrathi or Hooghly branch of the Ganges and there offer *srāddha* for the spirits of their ancestors.

Khardah. Khardah is a favourite place of pilgrimage for Vaishnavas, as it was the home of Nityānanda, one of the chief disciples of Chaitanya. They visit it more especially during the Rāsh and Phul Dol festivals in the months of Kārtik and Baisākh, respectively; the pilgrims include many prostitutes and other members of the lower classes of Calcutta.

Palābāri Jaynagar. The shrine of Rādhāballabh at Jaynagar is visited by a large number of pilgrims during the Dol festival in the month of Phālgun. Close to the temple is a *kadamba* tree which during the festival is said to yield one flower in honour of the god Krishna, whose favourite flower it is; this is looked upon as a supernatural phenomenon, for the rainy months of July and August are the season in which the plant flowers.

Ghāzi Sāheb's *Melā*. The *melā* of Ghāzi Sāheb is a religious gathering held in the latter part of May or in the early part of June at Bāsra, close to the station of that name on the Canning branch of the Eastern Bengal State Railway. On this occasion the votaries, both Hindu and Musalmān, visit the shrine of Ghāzi Sāheb (whose legend has already been given), and make offerings of sweetmeats, goats (in the case of Hindus) and fowls (in the case of Musalmāns) in order to obtain immunity from the attacks of tigers, or relief from sickness and disease.

Mānik Pīr's *Melā*. A similar religious gathering takes place at Jādabpur, close to the station of the same name, in honour of Mānik Pīr This attracts few but Musalmāns, who offer fowls, which they cook and eat on the spot.

CHRISTIANS. Of the 16,027 Christians in the district, 2,348 are Europeans or members of allied races, such as Australians or Americans, 571 are Anglo-Indians (the designation prescribed by the Government of India for the community generally known as Eurasians), and 13,108 are Indians. There are more Indian Christians than in any other district of Bengal, while the number of Europeans is exceeded only in Darjeeling. Of the former, 4,774 are members of the Anglican Communion, 2,785 are Baptists, 1,815 are Congregationalists, 565 are Methodists and 2,962 are Roman Catholics. Missionary effort dates back to the early part of the nineteenth century and has met with most success among the lower castes in the south of the district. The following is an account of the chief missions at work.

Society for the Propagation of the Gospel. The Society for the Propagation of the Gospel, which is a Church of England mission, started work in the district in 1823, when the Rev. W. Morton took over charge of seven schools, which had been opened three years before by the Society for Promoting Christian Knowledge, at Tollygunge, Ballygunge, Bhawānipur, Kālighāt, Puturi, Garia, and Birel. About the same time a school at Bāruipur, which had been started in 1820 by Mr. Plowden, the Salt Agent, was transferred to the charge of the Society. In 1833 Bāruipur was made the headquarters of a separate branch of the mission, with a European missionary in charge; and in 1837 a temple of Siva was presented to the mission by two converts of the village of Sajinabāria and converted into a chapel. In the latter year all the villagers of Bereali in Magra Hāt renounced caste and sought Christian instruction. A fierce persecution was raised against them by a Musalmān zamīndār, and, to prevent their eviction, the Society purchased the hamlet and so secured the foundation of a station at Magra Hāt. *Pucca* churches were built and consecrated at Jhānjra in 1844 and at Bāruipur and Māgra Hāt in 1846. At this time there were 1,443 converts and catechumens in the Bāruipur-Magra Hāt district, extending for 40 miles in a direct line from Altābāria in the north to Khāri in the south and containing 54 villages; native readers were stationed in all the principal villages. In 1853 the baptized converts numbered 1,031 and the catechumens 609, in the Tollygunge-Jhānjra district, where 20 years previously there had been only 66 baptized converts. The rapid progress of Christianity aroused opposition, and even persecution. On one occasion two European missionaries were besieged for two hours in the chapel at Andarmānik by a gang of Hindus armed with clubs, who were led on by an apostate Christian. On another occasion, when a Brāhman had been converted, the mission house was beset for two days by a mob of Hindus instigated by the local zamīndār, and the huts of several converts were burnt down at night, attempts to set the mission house on fire having failed.

The two chief missionaries in the first half of the nineteenth century were the Rev. D. Jones, who worked first as catechist and then as priest from 1829 to 1853, and the Rev. C. E. Driberg (1832—71). The years that followed the death of Mr. Driberg were years of depression. The staff was undermanned; supervision was insufficient; there were many secessions to other Christian bodies. There was a marked improvement after 1878, when Miss Angelina Hoare came to the district and took up the education of women, opening girls' schools, etc.; her work is now carried on by the Community of St. John the Baptist, commonly known as the Clewer Sisters. Another body which has done much for the spiritual life of the district is the Oxford Mission to Calcutta.

The work carried on by the Society is mainly pastoral, *i e.*, it works among those who are already Christians. Work is carried on in 5 districts, viz., Bāruipur, Magra Hāt, Tollygunge, Khāri and Canning, and from two centres, viz., Tollygunge and Magra Hāt, each of which is supervised by an Indian clergyman. In the Tollygunge centre there is a *pucca* church at Jhānjra and there are *kutcha* chapels at Jayadergot, Rāghabpur, Hogalkānia, Sālpukur, Pānākua, Kharibāria, Kālīcharanpur, Bethbāri, Balarāmpur, Cheyāri, Tollygunge, Kāorāpukur and Baddipur. In the Bāruipur centre there are two *pucca* churches, one at Bāruipur and the other at Magra Hāt, and 17 *kutcha* chapels situated at the following places:—Lakhīkāntapur, Dhānghāta, Khāri, Bāmanābād, Malayapur, Sālkia, Mākāltala, Ban Magra, Kālīpur, Andarmānik, Canning, Basanti, Rāmkrishnapur, Kola Hāzra, Tangrākhāli, Phulbāri and Bokultala. The Society maintains a number of primary day schools for boys, from which promising students are sent to the boarding house attached to the Oxford Mission Industrial School at Ballygunge. Promising pupils from girls' schools are sent to the Diocesan High School for Girls at Pipalpati, in Bhawānipur, which is under the Clewer Sisters. For the training of readers there is a vernacular readers' class at Jhānjra. Candidates for ordination are sent to Bishop's College at Ballygunge.

The institution called the Community of St. John the Baptist, or popularly the Clewer Sisters, is attached to the

Church of England. Its property is held in trust by the Bishop and Archdeacon of Calcutta, and the centre from which it works is the Diocesan Mission House in Ballygunge. Its sphere is chiefly educational, village schools for girls being kept up and regularly visited. This work, as stated above, was begun by Miss Angelina Hoare, and on her resignation was transferred by the Bishop to the Clewer Sisters.

Church Missionary Society. The Church Missionary Society maintains an orphanage at Agarpāra, which was founded in 1837 by Mrs. Wilson and handed over to the Society some 20 years later. At Barnagore there is a small congregation, composed mainly of women converts of the Church of England Zenāna Mission. Another congregation at Kristapur near Dum-Dum is composed of Tiyar fishermen. The inception of the mission in this village is due to the fishermen hearing street preachers in Calcutta when they came to sell their fish, and it began in a small way about the year 1850. The village contains a church built in 1871 and small schools for boys and girls. Another station has been established at Thākurpukur with a small girls' school, and *pāthsālās* for boys. There are also small communities of Indian Christians at Akra, where there is a chapel, and at Diamond Harbour.

Church of England Zenāna Mission. A branch of the Church of England Zenāna Mission was started in 1892 in the Barnagore thāna, where the ladies of the Mission teach some 400 children in village schools. The Mission also maintains an industrial school for women who have become converts to Christianity, so that they can earn a living by needlework and by making jams, chutnies, curry powders, etc.

Roman Catholic Mission. The Church of Rome carries on missionary work in the Sundarbans and its neighbourhood. It established a Sundarbans mission in the year 1868, when the first mud chapel was erected in the village of Kaikhāli, 16 miles south of Calcutta. Five years later it extended its operations in a southerly direction to the country lying between the Hooghly and Mātla rivers; and in 1876 it moved on further to the south-west. Eventually two central stations were established, viz., Morāpāi and Rāghabpur, to which some 24 villages were successively attached. In each of these two stations two priests reside. Those at Morāpāi visit 16 stations and out-stations, and there are central schools both at this place and at Rāghabpur. Rāghabpur possesses, in addition to two schools for boys and girls, a fine church, with seven mud chapels.

Baptist Mission. The Baptist Missionary Society began its labours in the 24-Parganas in 1827, in which year its first native church was established at the village of Narsidachak; this was the outcome of its propaganda in Calcutta, to which the villagers came on business. In a few months it extended its work to North and South Lakhīkāntapur, Rasulmahmud, Boālkhāli and Māchkhāli. A sub-station was opened at Khāri near the Sundarbans in 1829; in 1844, work was taken up at Vishnupur, on the Diamond Harbour Road, and in the neighbouring villages; in 1868 another station was started at Janjālia, and work was also opened up in the Mātla district. Other village stations have been established as converts multiplied, more recent additions being those at Bāghmāri, Hārbhāngi, Basanti, Chordākāthi, and Tangrākhāli. There are three chief centres from which the mission work is carried on, viz., Vishnupur, Lakhīkāntapur and Mātla, which are situated some 20 miles from one another. Primary day schools have been maintained ever since the foundation of the mission, and a flourishing boarding school has been in existence at Vishnupur for 30 years.

London Mission

The London Missionary Society is an undenominational body, founded by Episcopalians, Presbyterians and Independents, which is now practically the representative of the Independents. It was founded in London in 1795 and established its first Bengal station at Chinsura in 1798. Its work in the 24-Parganas began in 1819, when it set up stations at Kidderpore and Kālīghāt. The following stations were subsequently added:—at Bhawānipur in 1823, at Kāorāpukur and other villages to the south in 1826, at Beliāhāti and other Sundarbans villages in 1844 and at Bāduria in 1875. The mission maintains two high schools and a first grade college at Bhawānipur, as well as elementary schools.

Other Missions.

Missionary work is also carried on by the Church of Scotland, the Wesleyan Mission, the American Methodist Church and the Women's Union Missionary Society. The Society last named was founded in 1860, and claims to be the oldest woman's foreign missionary organization in America; it works among women and children and is undenominational.

CASTES AND TRIBES.

The following statement shows the Hindu castes and Musalmān groups that have a numerical strength of 25,000 and over according to the census of 1911.

Name.		No.	Name.		No.
Sheikh	...	675,082	Tiyar	...	64,058
Pod	...	333,747	Kāyasth	...	41,176
Kaibartta	...	244,515	Muchi	...	40,517
Ajlāf	...	119,364	Jolāha	...	32,996
Bāgdi	...	99,425	Nāpit	...	29,733
Brāhman	...	91,003	Baishnab	...	28,283
Goālā	...	74,229	Namasudra	...	26,874
Kāora	...	64,483			

The Sheikhs constitute by far the largest Musalmān group, accounting for three-fourths of the total number of Musalmāns in the district. There is an idea among the Musalmāns that they must necessarily all belong to one or other of the four classes of Saiyad, Pathān, Mughal and Sheikh, and the designation Sheikh is accordingly adopted by members of the various functional groups, and also by new converts, who are desirous of hiding their real origin. As pointed out by Mr. Gait, in the Bengal Census Report of 1901, a well-to-do man of a functional group, say a Jolāha, will discard the designation of Jolāha, and call himself a Sheikh, assuming a more respectable name. Thus, Chanu Miyān will become first Chainuddin, then Muhammad Chainuddin and finally Maulvi Muhammad Chainuddin Ahmad. This ascent of the social ladder is pithily described and satirized in the proverb—"Last year I was a Jolāha; this year I am a Sheikh; next year, if prices rise, I shall be a Saiyad." It is this tendency which is largely responsible for the addition of 101,000, or 17½ per cent., to the number of Sheikhs since 1901.

Musalmān groups.

The term Ajlāf, which in Bengali is corrupted into Atrāp, means "the lower classes", and is a generic designation applied to converts and various functional groups such as the Dhunia, or cotton cleaner, the Kulu or oil-presser, the Hajjām or barber, the Kunjra or vegetable seller, the Darzi or tailor, etc. The Jolāhas are weavers by hereditary occupation, but a large proportion have abandoned their looms, and engage in manifold other callings, *e.g.*, they are cultivators, butchers, mill-hands, shopkeepers, cartmen, etc. The name Jolāha is disliked by them because it is proverbially used as a synonym for a fool. They, therefore, call themselves Momin or Sheikh Momin, or even Sheikh, the two latter names being intended to substantiate their claim to recognition as Sheikhs.

Hindu castes.

It is not proposed to give an account of the numerous Hindu castes, for which the reader is referred to ethnological works, such as Risley's *Tribes and Castes of Bengal*. A brief mention will, however, be made of the Pods and Kāorās, castes which are more or less peculiar to the district. It may also be explained that Namasudra is the name now recognized for the caste formerly known as Chandāl, and that the Kaibarttas are divided into two main classes, viz., the Chāsi and Jaliya. The Chāsi Kaibarttas, who are beginning to discard the old nomenclature and call themselves Māhishya, are cultivators and superior in status to the Jaliya Kaibarttas, who are fishermen. At the census of 1911 only 4,015 Kaibarttas returned themselves as Jaliya Kaibarttas.

while the number of those who stated that they were Chāsi Kaibarttas or Māhishyas was 240,487.

Pods. More than half the Pods of Bengal are residents of the 24-Parganas, practically all the remainder being found in adjoining districts, viz., Khulna, Midnapore, Howrah, Jessore and Hooghly. There are four main sub-castes, viz., Chāsi Pod, Mecho Pod, Tānti Pod and Bhāsa or Dhamna Pod. The differentiation between them appears to have had its origin in differences of occupation, as indeed may be gathered from the names. Each sub-caste has a separate *panchāyat* for the adjudication of cases concerned with offences against caste and social laws, but all have the same class of priest. The Chāsi Pods, who live by cultivation, are recognized as having a superior status, and claim to be of Kshattriya origin. They accordingly call themselves Brātya Kshattriya or Padma Rāj and deny all connection with the Mecho Pods, who live by fishing; occasionally, however, they will marry Mecho Pod girls, though they will not allow their daughters to marry Mecho Pod men.*

The Bhāsa Pods appear to be a comparatively recent accretion from outside. The name Bhāsa means "floating," and tradition has it that the Bhāsas were washed over to the 24-Parganas from Hijili and other places in Midnapore in the cyclones of 1824 and 1834. Originally the name seems to have been applied generally to such immigrants without special reference to any caste, and the application of it to a Pod sub-caste is apparently a modern innovation. The Bhāsas are found in considerable numbers in Sāgar Island, where the original immigrants settled. They allow widow marriage and are looked down upon on that account by the other sub-castes, who will not smoke from the same hookāh as these laxer brethren.

Kāoras. More than half of the Kāoras in Bengal are found in the 24-Parganas, practically all the remainder being inhabitants of the adjoining districts of Hooghly, Howrah, Midnapore, Jessore and Khulna. They are an extremely low caste—so low indeed that they may not enter even the courtyard of a temple and may not take part in any religious ceremony, except as drummers. Their touch defiles Brāhmans and Kāyasthas, and the Dhoba and Nāpit will not serve them. There are three sub-castes in this district, viz., Chhāhi, Dāi Kāora or Hāri Kāora and Bāburji. The Chhāhi work as cultivators, chaukidārs, labourers and drummers; they claim superiority to the other sub-castes and will not eat, drink or smoke with them. The Bāburji work as cooks in the

---

* Bengal Census Report of 1901, p. 382.

houses of Europeans and the Dâi or Hāri Kaorās as *palki*-bearers and drummers. The latter also keep and sell pigs, while their women are employed as midwives. The priests of the Kaorās are generally degraded Brāhmans, who are not much better than themselves. These Brāhmans are called Pandit—a title also given to men of the caste who officiate as priests in remote places where Brāhmans are not available; the insignia of these Kāorā priests is a copper ring worn on the fore-arm.

Caste government. The internal affairs of the castes and sub-castes are regulated by meetings of the more influential members, and among the lower castes there are recognized headmen to whom complaints are preferred and with whom information is laid of suspected offences against caste laws. These headmen are generally called Mandals or Samājpatis, but the Pod headmen are known as Sardārs. They hold office by hereditary right, but there is a tendency for their places to be taken by parvenus who have amassed wealth and consequently acquired influence among their neighbours. Their jurisdiction is usually limited to the community resident in the *pāra* or village, but in some cases extends to a group of villages. The meetings are held as a rule on the occasion of a marriage or funeral, when there will naturally be a gathering of the caste men, and the discussion is post-prandial. Heinous offences are visited with excommunication, which may be either temporary or permanent, and fines are inflicted for venial delinquencies. The fines when paid are generally credited to the *lārwāri* fund, which defrays the expense of village *pūjas*, or to the village Hari Sabha. Sometimes, however, the proceeds are spent on caste feasts or in feeding Brāhmans, and occasionally the culprit is made to undertake some work of public utility, such as the excavation of a tank, the repairs of a temple, etc.

CHAPTER IV.

PUBLIC HEALTH.

HEALTHY AND UNHEALTHY AREAS.

In the Bengal Census Report of 1901 the thanas of the district were grouped, according to their physical features, in five blocks, of which the following description was given :—

"(1) Six riparian thanas lying along the bank of the Hooghly, viz., Naihāti, Nawābganj, Barrackpore, Khardah, Barnagore and Budge-Budge. These thanas, on the whole, are more healthy than those further inland, being higher, and enjoying a better supply of drinking water, which, in some cases, is filtered by the management of the mills.

"(2) The north and central thanas of Hābra, Deganga, Bārāsat, Dum-Dum and Tollygunge. The drinking water is here very bad, being derived mainly from tanks polluted by surface drainage ; the drainage channels are blocked and there are numerous swamps, and the homesteads are surrounded by dense jungle. Malaria is very prevalent.

"(3) The eastern thanas of Bāduria and Basirhāt. The inhabitants are, for the most part, sturdy Muhammadans ; the country is now healthy, and the main crop is jute, which yields a handsome profit to the cultivators.

"(4) The southern thanas, viz., Haroa, Bhāngar, Sonārpur, Bāruipur, Vishnupur, Falta, Diamond Harbour and Magra Hāt. These thanas are salubrious, owing to better drainage, the comparative absence of noxious undergrowth, and the sea-breeze that blows almost continuously during the south-west monsoon.

"(5) The Sundarbans thanas, viz., Hasanābād, Canning or Mātla, Jaynagar, Mathurāpur and Kulpi. Cultivation is here spreading rapidly, and reclamation is extending southwards."

The relative healthiness of these five blocks was inquired into by the Bengal Drainage Committee in 1907, and the following conclusions were arrived at : "From the description given in the Census Report, the areas of most malaria might be looked for in the second division, and the remainder might be expected to be comparatively healthy, but, upon the basis of the vital statistics, it will be seen that it is the thanas of the first class which are the most feverish. The explanation possibly lies in the fact that the Census

Report had more in mind the municipal portions of these thanas, along the bank of the river; the statistics quoted are of rural areas, away from the river, the conditions of which approximate more to those of the country further east. The annual average statistics of mortality, from all causes and from fever, place the different thanas in approximately the same relative order, but Tollygunge and Basīrhāt are rather more feverish, and Diamond Harbour considerably less so, than the figure of total death-rates would show. Upon the basis of the district fever rate (18·3 per mille), we may, perhaps, class rates of 25 and over as unhealthy, and of 15 and under as the reverse. If so, the specially unheathy thanas in the 24-Parganas are Dum-Dum, Khardah, Barrackpore, Naihāti and Nawābganj, on the east bank of the Hooghly north of Calcutta, and Deganga and Hābra, adjoining them further inland to the east. The healthy thanas are Bhāngar, Mātla, Diamond Harbour and Budge-Budge, all, as might be expected, to the south towards the sea. The conclusions to be drawn from the census fluctuations are obscured by the shifting of the population, due to the opening of new industrial centres, etc.

"In the matter of public opinion, we consider the tendency is to class the district as more unhealthy than it is. This is intelligible; grievances can be easily represented in Calcutta, and the standard of comparison is, perhaps, higher than in less advanced districts. Apart from individual villages, the District Magistrate names the thanas of Bārāset, Hābra, Deganga, Dum-Dum, Barrackpore, Nawābganj, Khardah and Naihāti as specially unhealthy, and the statistics bear out this view; but, in the case of Haroa, Falta, Jaynagar, Mathurāpur and Magra Hāt, which are also mentioned, this is scarcely the case for the area of the whole thana, but the reference is apparently to particular portions only.

"The only recent detailed local inquiry in this district was made in 1900, by Captain Rogers, when urban spleen rates of 55 (Gobardānga), 52 (Basīrhāt), 56 (South Barrackpore) and 68 (North Dum-Dum) were taken, among others, all pointing to the prevalence of malaria. To sum up the local conditions:—

(a) The district as a whole is not abnormally unhealthy, though some portions of it return high rates of mortality.

(b) Similarly, it is not, as a whole, specially malarious.

(c) The noticeably malarious thanas are those of Dum-Dum, Khardah, Barrackpore, Naihāti, Nawābganj, Deganga and Hābra.

(d) The least malarious areas are the Bhāngar, Mātla, Diamond Harbour and Budge-Budge thanas."

These conclusions are corroborated, to a large extent, by Major A. B. Fry, I.M.S., Special Deputy Sanitary Commissioner, Malaria Research, Bengal, who, in his *First Report on Malaria in Bengal* (1912), has published a chart showing the distribution of malarial intensity, as evidenced by mortality returns, from which it appears that the thanas may be classed in five blocks according to the incidence of malaria. They are, in order of intensity, (1) Naihāti and Dum-Dum, (2) Barrackpore and Khardah, (3) Bāduria and Hābra, (4) Bārāset, Basirhāt, Budge-Budge, Noapāra, Tollygunge and Vishnupur, and (5) Debipur, Diamond Harbour, Magra, Jaynagar, Bāruipur, Sonārpur, Bhāngar, Deganga, Haroa, Hasanābād, Alipore and Behāla.

VITAL STATISTICS. Both the birth-rate and the death-rate are low, compared with other Bengal districts, the average birth-rate for the ten years 1893-1902 being 27 per mille, while the average death-rate was only 24 per mille. The ratios have since risen, but are still comparatively low, the averages for the ten years ending in 1910 being 32 and 27 per mille, respectively. The explanation is that there is a preponderance of males, and a deficiency of females of the child-bearing age, the natural consequence of which is a low birth-rate; and a low death-rate is a corollary to a low birth-rate more especially as infantile mortality is responsible for a large proportion of the deaths.

As in other districts, the majority of the deaths are ascribed to the general head of fever; the average reported fever mortality since 1892 (when the present system of reporting and recording vital occurrences was introduced), is 17 per mille. Inquiries made by qualified medical officers elsewhere have shown that roughly one-third of the reported fever deaths are really due to malaria; and this conclusion is confirmed, so far as the 24-Parganas are concerned, by a series of observations made in the Dum-Dum thana during 1911-12 by Assistant Surgeon Babu Mithiles Chandra Ghosh. The report of that officer, who verified over 300 cases, is instructive, as showing the diseases which are most prevalent, and also the degree of accuracy attained in the classification of deaths by the agency responsible for the returns. "Of 32 cases returned as cholera, all were due to that disease. Of 31 cases returned as dysentery, 26 were correct. Of 219 cases returned as fever, 143, so far as I can make out, had no connection with malaria. Therefore, 34·7 per cent. of the deaths from fever are due to malaria. Many cases of dysentery, which terminate often with fever, were put down as fever. There were 29 cases which I have classed as enteric fever, as the history given showed that all had continued fever with diarrhœa and complete prostration. There were 10 cases of very old people who do not seem to have had any particular illness, but died of old age and debility. Measles, septicæmia, tuberculosis, phthisis and pneumonia account for many deaths. Tetanus neonatorum is described by the villagers as the possession of a child by a she-devil, which causes the convulsions. It is due to lack of cleanliness on the part of the ignorant *dhai*, who is a very low caste woman. The type of malaria present here is the chronic endemic form. Of the 76 deaths which I have classed as malarial, 62 were chronic cases of long duration. I have found eight cases of *kála azār* amongst children. I diagnosed them on clinical grounds, and in no case made spleen puncture, but in each case I took several blood films on Major Donovan's method, and in one case I found a parasite. These cases seem to be sporadic, and no two cases were in the same family."

Cholera has a tendency to become epidemic twice a year, viz., at the beginning of the hot weather, and at the end of the rains. This scourge, however, rarely assumes the proportions of virulent intensity which it does in some districts. The worst epidemic in recent years occurred in 1907, causing 14,500 deaths, or 7 per mille of the population. Small-pox occurs on a small scale every succeeding spring, but rarely assumes a seriously epidemic form: the highest mortality due to it of late years was returned in 1909, when the total number of deaths was only 2,000. Epidemic fever, similar to that raging in Burdwān and Hooghly, broke out in the Bārāset subdivision in 1862, and continued for some years committing great havoc, but no statistics are available showing the proportion of the population affected or the rate of mortality. EPIDEMIC DISEASES.

The following account of the fevers found in the district was contributed a few years ago by Lieut.-Colonel Harold Brown, I.M.S., when Civil Surgeon of the 24-Parganas:— TYPES OF FEVER.

"The fevers of the district may be roughly divided into two classes, viz., the malarial and the non-malarial.

"*Malarial Fevers.*—These are caused by the presence in the blood of a minute amœba, known as the *plasmodium malariæ*, and are introduced into the system by the agency of various species of malaria-carrying mosquitoes, the *anophelidæ*. Malarial fevers are divided into three classes, each having its own special form of parasite: these are the benign tertian, the benign quartan, and the malignant infection.

"*The Benign Tertian.*—The parasites in this form have a cycle of 48 hours, and hence cause a fever that recurs every second day (tertian ague). This is the mildest of all the forms of malarial fever, as well as the commonest, and occurs from July till March, being commonest in the cold months. It is readily amenable to treatment by quinine and, even if untreated, gradually wears itself out, this form of fever seldom being directly fatal, though, if neglected, it eventuates in enlargement of the spleen. There is often a double infection by this parasite, quotidian ague resulting.

"The other benign parasite is that which has a cycle of 72 hours, causing quartan ague: this is a decidedly rare form of fever, though one meets with a certain number of cases every year. It is also intractable, unless carefully treated with quinine. As usually met with, the fever produces a chart in which there is a rise of temperature every third day, there being a fever-free interval of 48 hours, but if there is a double or treble infection, the type will vary accordingly, so that even a quotidian ague can be produced.

"*Malignant Infection.*—There are several varieties of this form, including a malignant quotidian, tertian and the so-called remittent, which is the result of repeated infection. These are the most important, and the most severe forms of malarial fever, producing the common type that was formerly known as 'malarial remittent', as distinguished from 'intermittent' fever. In these cases, the temperature seldom falls to the normal point, but the chart is generally irregular. Untreated, they run a course of three weeks or more, when, if not fatal, they tend to develop an intermittent temperature and slowly decline, but are very apt to relapse. It is to this form that the great mortality from 'fevers' is due, and, even when taken in hand early, many cases are very rebellious to treatment. Many of the cases resemble 'typhoid' fever, but a careful examination of the blood will prevent the occurrence of a mistake, as the parasites will be found in the former, but not in the latter. Not many years ago the name 'typho-malarial' was applied to certain cases of long continued fever, with symptoms of depression like those seen in typhoid: some of them were true malaria, others true typhoid, and nowadays a hybrid form of the disease is not believed in. There are also a quotidian and a tertian type of malignant infection, which are far more serious than similar forms of fever, due to infection by the benign tertian parasite.

"The vast majority of cases of 'remittent' fever, due to the presence of the malarial parasites, can be cured by quinine. In some cases, however, when the infection is very acute, and coma is present, owing to an accumulation of the parasites in the vessels of the brain, it is useless to administer the drug by the mouth, and it should be administered hypodermically, in doses of 10—15 grains, either under the skin, or deep into the muscles of the buttock. A great many cases are still lost by the practitioners withholding quinine until an intermission or remission of the fever occurs. In the old days, it was taught that quinine was dangerous if given during the height of the fever, so the medical attendant waited for a decline of the temperature. Very often there was no decline, and the patient died from the bad expectant treatment, but nowadays we know that, the higher the temperature, the larger is the dose of quinine that is necessary, and the drug is injected without a moment's delay. In some forms, there is great irritability of the stomach, with sickness, so that it is impossible to give quinine by the mouth; here, again, the drug must be injected. Another form of remittent fever is that accompanied by great coldness of the surface of the body, with collapse, though the thermometer indicates a temperature of 104°, or higher, in the mouth or rectum. These cases cause great anxiety, and are often fatal. In other cases, again, the attack of fever is associated with dysentery, but here quinine, and not the treatment of ordinary dysentery, is indicated.

"*Non-malarial Fevers.*—Of the non-malarial fevers occurring in the district, the following are the most interesting. 'Cachectic fever,' formerly considered to be malarial and known as 'malarial cachexia,' but now regarded as due to an infection by the Leishman-Donovan body. This has recently been shown by Major Rogers to be a stage in the development of a trypanosome. This fever is very common, and is accompanied by great enlargement of the spleen, dropsy of the face, extremities and body, and profound anæmia. It is certainly not malarial, and is responsible for a great many deaths annually. The parasite is found chiefly in the spleen, where it can be obtained in large numbers, but, as a rule, none can be found in the peripheral blood. Of recent years, a fever has been prevalent in Calcutta and the suburbs, which has been called the 'seven days' fever.' It is not malarial, and at first was considered to be a mild form of influenza. The opinion has been expressed that it is a mild form of dengue, but, in my opinion, this is not the case, as it is very feebly infectious, whereas dengue is one of the most infectious diseases known. The peculiarity of the fever is its regular course, lasting 6 to 7 days.

"Typhoid fever, which, at one time, was supposed not to be found in the natives of India, is really a very common disease in the suburbs of Calcutta, and is responsible for a considerable proportion of the mortality from fevers. It was formerly confused with malarial remittent, but, as the two can be differentiated with a little trouble, it is now evident that typhoid fever is much more prevalent than most people think. It attacks young adults and children chiefly, is not controlled by quinine, and tends to run a course of three or four weeks or, if there are complications, longer. The diagnosis is arrived at by the failure to find malarial parasites in the blood, and by a positive reaction of the blood serum to a fresh culture of the *bacillus typhorus*, a reaction known as 'Vidal's.' Typhoid fever is less common in the mofussil than in the suburbs of Calcutta, but cases are frequently met with all over the district

"Of late years cases of cerebro-spinal meningitis have been observed from time to time, and, in 1901, when investigating an epidemic in the Emigration Depôts at Garden Reach, I found that the disease was endemic in Calcutta and the suburbs, and was often mistaken for plague. It is one of the most fatal of fevers, the mortality varying, according to the type, from 30 to 100 per cent. the average mortality being over 60 per cent. Occurring among the general population, it is generally sporadic, but where human beings are collected in large numbers, as in the Alipore Central Jail and the Emigration Depôts, it sometimes occurs in epidemic form, and I have known 40 cases in a single Emigration Depôt in two months. The disease is due to a specific bacillus, which can be obtained in the cerebro-spinal fluid in every case, and, as before mentioned, the mortality is very great."

OTHER DISEASES.

Dysentery and diarrhœa account for a considerable number of deaths every year; they are most prevalent in May, June and July, and are least in evidence during November, December and January. Tubercle of the lungs is very common: in fact, it is one of the commonest affections for which patients seek relief at the hospitals. The causes of its prevalence are, briefly, poverty and consequent inability to procure adequate and suitable food, filthy and insanitary environments, insanitary occupations, neglect to appreciate the gravity of early symptoms, and, lastly, poor physique and diminished power of resistance to disease. Asthma is frequently seen, and is very rebellious to treatment. Glycosuria is common; it is met with not only among middle-aged men of the educated classes, but also among young adults of the lower orders. Venereal diseases are rife. Syphilis, gonorrhœa and soft chancre are seen daily and contribute a large proportion of the out-door patients at the hospitals and dispensaries. Elephantiasis is common, but large tumours due to this cause are not common nowadays, as the subjects seek relief by operation earlier than used to be the case. Goitre and stone are comparatively rare, and the number of operations for vesical calculus is insignificant. Bright's disease is not uncommon. Valvular affections of the heart are frequently seen. Malignant tumours are not uncommon. Hydrocele and nakra are of very frequent occurrence. Plague has been in evidence during the last decade, but outside the environs of Calcutta there have been only isolated cases, and there has not been an epidemic worthy of the name. Cases of cerebro-spinal meningitis, which, as already stated, is apt to break out at the emigration depôts and in the suburbs of Calcutta, are sometimes mistaken for plague.

WATER-SUPPLY. The suburbs of Calcutta receive their supply of drinking water from the Calcutta water-works; this is a good pure water, which is distributed as far south as Garden Reach In the north of the suburban area Cossipur, Chitpur and Māniktala are supplied, and in the south and south-east Alipore, Ballygunge, Tollygunge, Behāla and Barisa. Wherever there are mills, there is a plentiful supply of good filtered water, which is taken from the river, pumped up to reservoirs and distributed by means of pipes. In this way the people living near the mills, from Barnagore up to Naihāti, receive a supply of drinking water equal to that obtained in Calcutta. Elsewhere they are dependent on the rivers and tanks, which are only too often polluted by surface drainage. Well water is unpopular, and is very little used. From Garia to Sāmukpota, along Tolly's Nullah, the water is so brackish as to be unfit for drinking, and the people are obliged to walk miles to obtain a supply for their households. In the Sundarbans also it is often impossible to obtain sweet potable water, but the people appear to become inured to brackish water and drink it without any injury to themselves.

HOSPITALS AND DISPENSARIES. In the returns published by the Inspector-General of Civil Hospitals in the annual report on charitable dispensaries in Bengal, 38 charitable dispensaries and hospitals are shown as in existence in the 24-Parganas. In addition to these, there are three large hospitals, which are grouped with the Calcutta medical institutions in the returns, as they are situated in the Added Area of Calcutta, but which come within the administrative limits of the district, viz., the Campbell Hospital at Sealdah, the Presidency General Hospital and the Sambhu Nāth Pandit Hospital at Bhāwanipur. The first two are purely Calcutta institutions and have no connection with the district of the 24-Parganas, except that the accident of their site brings them within its area. The Sambhu Nāth Pandit Hospital is on a different footing, for it is directly under the Civil Surgeon of the district, who is *ex-officio* Superintendent of the hospital, and it is practically the head-quarters hospital of the district It is, therefore, included in the scope of this chapter (whereas the other two are excluded), and is entered in the following table of hospitals and dispensaries :—

Hospital or Dispensary.	No. of beds (1911)	DAILY AVERAGE OF PATIENTS (1911).	
		Indoor.	Outdoor.
Bāduria	...	...	126
Bāraset	10	4	46
Barisa	6	4	31
Barrackpore (Bhola Nāth Bose's) ...	20	10	96
Ditto (Cantonment)	19	7	57
Bāruipur	...	...	97
Basirhāt	6	1	40
Belpukur	...	...	28
Bhātpāra (Female)	...	...	36
Birāti (North Dum-Dum)	...	...	24
Budge-Budge	6	3	58
Canning Town	...	...	29
Chetla (Prince Ghulām Muhammad's)	...	...	44
Chitpur	...	...	139
Cossipur (North Suburban Hospital)	40	31	78
Diamond Harbour	18	6	51
Dum-Dum (Cantonment)	8	1	16
Do. (South)	...	...	24
Garden Reach	...	...	53
Gārulia	3	...	31
Hābra	...	...	15
Hālisahar	...	...	22
Harināvi	...	...	47
Haroa	...	...	51
Jaynagar (Allen's)	...	...	21
Kākdip	...	...	21
Khardaha (South Barrackpore)	...	...	21
Kidderpore (Municipal)	...	...	63
Kolsur	...	...	45
Magra Hāt	...	...	48
Māniktala (Narkaldānga)	...	...	57
Manirāmpur (North Barrackpore)	...	...	19
Naihāti	...	...	33
Palta (North Barrackpore)	...	...	19
Russa (Prince Ghulām Muhammad's)	22	9	55
Sambhu Nāth Pandit Hospital	68	65	213
Tāki (Tāra Sankar Chaudhuri's)	...	...	94
Tentulia (Bil Balli)	...	...	66
Ultadānga	...	...	18
Vishnupur (Stevenson-Moore's)	...	...	29

Three of the dispensaries entered in this table are not under the Civil Surgeon, viz., the Kidderpore Municipal Dispensary, which is under the Calcutta Corporation and its Health Officer, and the Cantonment Dispensaries at Barrackpore and Dum-Dum, which are under officers of the Royal Army Medical Corps, who act as Civil Medical Officers of those stations. There are also some dispensaries which are not strictly speaking charitable dispensaries, viz., the Alipore Police Case Hospital, the Royal

Indian Marine Dockyard Dispensary at Kidderpore and the Eastern Bengal State Railway dispensaries at Sealdāh, Kānchrāpāra and Chitpur, which treat only policemen and dockyard and railway employés, respectively.

The following account of individual hospitals and dispensaries is furnished from notes, prepared for a medical history of the district, which were supplied by from the Civil Surgeon's office :—

The Sambhu Nāth Pandit Hospital.—The fore-runner of the Sambhu Nāth Pandit Hospital was the Bhawānipur Dispensary, which was opened in 1840. It was entirely maintained by Government up to 1871, when it received a grant of Rs. 21,445 from the Sambhu Nāth Memorial fund, on condition that the dispensary should be named after Sambhu Nāth Pandit. Of this sum, Rs. 15,645 was spent on the acquisition of a site on the east and near the north end of Russa Road and on building; the balance (Rs. 5,800) was invested in Government securities. This dispensary treated out-patients only, and was finally closed in 1896, on the opening of the new hospital.

In 1890-91 a special committee of the Calcutta Corporation recommended the building of a new hospital in Russa Road, for the benefit of the southern suburbs, and this proposal was supported by the Suburban Improvement Committee. At that time however, Government was not prepared to help, and the project fell through. It was revived by a committee which was appointed, by the Bengal Government in 1893 to consider the medical needs of Calcutta. One of the recommendations it made was that a hospital should be established in Bhawānipur, with 50 beds for ordinary in-patients, six beds for infectious cases and five for cholera cases. This proposal was taken up by the Calcutta Corporation, which in 1894 passed the following resolution :— "That the Commissioners were prepared to give a site in Elgin Road, south of the water-works, for the new hospital; that they would merge the Sambhu Nāth Pandit Dispensary in the new hospital, provided that the outdoor dispensary continued to bear that name; that they would hand over to Government the value of that dispensary, after sale, and the endowment of Rs. 5,800; that they would make an annual contribution of Rs. 5,000 to the new hospital; and that the property should be vested in a Board of Governors." The new hospital was accordingly built on the north side of Elgin Road, Bhawānipur. The site given by the Corporation was valued at Rs. 64,366, and the cost of building the hospital (including some additions, a portico, a *post mortem* house, etc., made during the two years subsequent to its opening) came to Rs. 1,41,963, which was paid by Government. Accordingly, a great deal more than half the cost of its erection, as well as more than half the cost of its maintenance, was borne by Government.

In 1898 the Corporation handed over to Government, for the extension of the hospital, a block of land situated on the north-east of the hospital grounds. on condition that the whole institution should be named the Sambhu Nāth Pandit Hospital. This block was utilized for the erection of an out-patient department, the quarters of the Civil Hospital Assistant, a contagious ward and a cholera ward.

The Civil Surgeon of the 24-Parganas is *ex-officio* Superintendent of the hospital, and its affairs are regulated by a Board of Governors, of which the Commissioner of the Presidency Division is President. There are five other members of the Board, of whom three are appointed by Government and two by the Calcutta Corporation. Its income in 1911 was Rs. 63,856, including a cash balance of Rs. 26,085. It received Rs. 26,591 from Government and Rs. 9,398 from the Calcutta Corporation, while Rs. 1,136 were realized from interest on investments. Private subscriptions amounted to only Rs. 551.

Alipore Police Case Hospital—Situated at Alipore between the Central Jail and the Magistrate's cutcherries. It was opened in 1852, and for 30 years was the Sadar dispensary of the district. It was made a police-case hospital in 1883, and is maintained entirely by Government. It treats both in-patients and out-patients, but chiefly the former. It is under the immediate charge of the Assistant to the Civil Surgeon of the 24-Parganas (an appointment sanctioned in 1885); the Civil Surgeon is expected to "generally supervise" his work, and usually visits the hospital about once a week.

Bāduria.—Opened in 1883. A *kutcha* building on a *pucca* plinth, with a deep thatched roof. It is maintained by the Bāduria Municipality with a contribution of Rs. 10 a month from the District Board; there are no private subscriptions. It treats out-patients only.

Bārāset.—Opened in 1854. It consists of two *pucca* buildings, one for in-patients and the other for out-patients, with a small contagious diseases ward and a cottage ward; there are 10 beds. It is maintained by the Bārāset Municipality, with a subvention of Rs. 10 a month from the District Board. It has an invested capital of Rs. 1,000, which in 1911 yielded Rs. 52. The dispensary is near the jail and three-quarters of a mile from the station (Eastern Bengal State Railway). Government keeps an Assistant

Surgeon at Bārāset, who is in medical charge of the subdivision, and receives Rs. 20 a month for the dispensary.

Barisa.—Opened in 1872. It is maintained by the South Suburban Municipality and receives a grant of Rs. 300 a year from the District Board; there are no private subscriptions. The building, which was damaged in the earthquake of 1897, is *pucca*, and has an upper storey which forms the quarters of the medical officer in charge; there are six beds for in-patients. It is situated on the west side of the Diamond Harbour Road just after the sixth milestone (from Government House) at the north end of Bāduria village.

Barrackpore—Bhola Nāth Bose's Dispensary.—This dispensary owes its existence to the liberality of the late Dr. Bhola Nāth Bose, for many years Civil Medical Officer of Farīdpur, who died in 1883, leaving his property in trust to Government after the death of his widow. On her death, in 1888, the whole property came under the Trust. It consists of Rs. 1,53,700, of which Rs. 1,49,700 are invested in Government securities, Rs. 3,000 in Bank of Bengal shares and Rs. 1,000 in Municipal debentures. A certain sum is paid to relatives of the testator, a grant is made for the support of the Mandālāi Dispensary in the district of Hooghly, and the balance falls to the share of the Bhola Nāth Dispensary. The amount derived from this source in 1911 was Rs. 2,080; Rs. 600 were contributed by the District Board, and Rs. 300 by the municipality, while Government made a grant of Rs. 1,574, and Rs. 3,408 were obtained from other sources.

The dispensary was opened in 1895, and, under orders issued by Government next year, was placed under the management of a committee composed of the Chairman of the District Board (as President), the Cantonment Magistrate, the Civil Surgeon of Barrackpore, and the Chairmen of North and South Barrackpore Municipalities: the Chairman of the Titāgarh Municipality was subsequently appointed a member of the committee. The Chairman of the District Board is *ex-officio* administrator of the fund, and the District Board practically manages the dispensary. The medical officer in immediate control is a Civil Assistant Surgeon. The buildings were erected by Government in 1894-95, at an estimated cost of Rs. 24,358, and were taken over by the District Engineer in 1895. They include (1) the main building, with three wards and 20 beds for in-patients, with an office, operation room and dispensary, (2) four small family quarters for patients, with separate cook-houses and latrines; and (3) a cholera ward, a pauper ward, a mortuary and quarters for the medical officer's compounder and servants. The dispensary stands on the east side of the Grand Trunk Road and is about one mile from the Barrackpore station.

The Barrackpore Cantonment Dispensary—Was originally opened in 1874, but was closed in 1880 on the cantonment subsidy being withdrawn. The present dispensary was opened in 1884; it is maintained chiefly by the cantonment funds. It is located in a detached building belonging to the Station Hospital, less than a mile from the railway station. This dispensary is not under the Civil Surgeon of the 24-Parganas, but sends in its returns through him to the Inspector-General of Civil Hospitals. It is directly under the Civil Medical Officer of Barrackpore, one of the officers of the Royal Army Medical Corps, who receives an allowance for performing the civil medical duties of the station.

Bāruipur.—Opened in 1883. It is supported by the Bāruipur Municipality, with a grant of Rs. 10 per month from the District Board and a few subscriptions. It is located in a *pucca* building erected in 1900-01, and treats out-patients only. It is situated on the Bāruipur-Jaynagar Road, about one mile from the Bāruipur station of the Eastern Bengal State Railway (Diamond Harbour branch), about 16 miles by road (*pucca*) from Alipore and about five miles from the Harināvi dispensary at Rājpur.

Basīrhāt.—Opened in 1867. It is maintained by the Basīrhāt Municipality, with a subvention of Rs. 10 per month from the District Board. It treats both in and out-patients, and contains six beds. The out-patient department is *pucca*, with quarters for the medical officer behind. The in-patient department has *pucca* walls and floor and a thatched roof. Government keeps a Civil Hospital Assistant at Basīrhāt, who is in medical charge of the subdivision, and receives an allowance of Rs. 10 a month from the dispensary funds.

Budge-Budge.—Opened in 1898. It is maintained by the municipality, and receives an annual grant from the Port Commissioners and subscriptions from some European firms.

Canning.—Opened in 1876. It is under the District Board, which took it over in 1899. In addition to the sums allotted for its support by the District Board, it receives annual grants from the Khās Mahāl Fund, the Port Canning Company and the Eastern Bengal State Railway. The dispensary has no building of its own, but is accommodated in a room of the inspection bungalow—a *pucca* building on a high plinth. It is about a quarter of a mile from the railway station, and 100 yards from the landing stage on the Mātla river. Canning is a small village, and patients are mostly drawn from outlying villages in the Sundarbans. The attendance must, therefore, always be small, unless Canning develops into a port and centre of trade. The dispensary, however, is the only one in the district within the Sundarbans line, and, even with its small attendance, distinctly meets a want. Out-patients only are treated.

Chetla (Prince Ghulām Muhammad's).—Opened in 1892. It is managed by the same committee as Prince Ghulām Muhammad's dispensary at Russa. This committee consists of the Collector as Chairman, the Treasury Officer as Secretary, the District Judge, the District Engineer, and the Civil Surgeon, *ex-officio*, a representative of the donor's family and two members appointed by the Calcutta Corporation. The dispensary was started with the balance of the interest of the Prince Ghulām Muhammad Fund available after paying for the upkeep of the Russa dispensary, and is now mainly supported by a grant made by the Calcutta Corporation. Out-patients only are treated. The building is a hired house with a small upper storey, which serves as the Medical Officer's quarters. Chetla is on the west of Tolly's Nullah, nearly a quarter of a mile from the bank, nearly opposite to Kālighāt and about 1½ mile from Alipore by a *pucca* road.

Chitpur.—Opened in 1893, and entirely maintained by the Cossipur-Chitpur Municipality. It treats out-patients only; there are no private subscribers. The building is an annexe to the Municipal office, which is situated on the west side of the Grand Trunk Road to Barrackpore, about half a mile north of the boundary of Calcutta City, and about six miles from Alipore.

Cossipur North Suburban Hospital.—Opened in 1866. It is managed by a committee, and is maintained partly by interest on investments, partly by contributions from various public bodies and partly by private subscriptions. It treats both in-patients and out-patients. The invested capital amounts to Rs. 33,500, of which Rs. 9,500 is a special endowment for the maintenance of a female ward, given by Babu Binod Lāl Ghose; the interest in 1911 was Rs. 2,160. The building is a two-storeyed one, containing four large wards, with ten beds in each. There are also male and female consulting rooms, a compounding room, an operation room and a store room. The medical officer has quarters in a two storeyed house, behind (east of) the hospital. The hospital is situated on the east side of the Cossipur Road, about one and-a-half miles north of Calcutta and about seven miles by road from Alipore. The medical officer is a Government Assistant Surgeon.

Diamond Harbour.—Opened in 1873, and made over to the District Board in 1892, by which body it is still maintained. It receives a grant of Rs. 300 a year from the Khās Mahāl Fund and of Rs. 180 from the Port Dues Fund. It treats both in-patients and out-patients and contains 18 beds. The building is *pucca*, with four rooms, one for female and two for male in-patients, the fourth being used as a consulting and compounding room. There is a bath-room with the usual furniture, so that one room can, if necessary, be used for the accommodation of Europeans. Occasionally sick Europeans are landed from ships lying at Diamond Harbour, but as a rule such cases would be transferred to a Calcutta hospital. The medical officer in charge is an Assistant Surgeon, who is in medical charge of the subdivision, and draws an allowance of Rs. 20 per month for the charge of the dispensary.

Dum-Dum Cantonment Hospital.—Opened in 1889 as a military hospital known as the Followers' Hospital. From 1889 to 1896 the Eastern Bengal State Railway contributed Rs. 25 a month to its support, but this grant was withdrawn at the end of 1896, after which the hospital, having no income or means of support, did practically no work until 1898, when the District Board came to the rescue with a grant of Rs. 10 (since increased to Rs. 20) a month. The dispensary is under the Civil Medical Officer of Dum-Dum, an officer of the Royal Army Medical Corps, who receives an allowance for performing the civil medical duties of the station.

Gārulia.—Opened by the North Barrackpore Municipality in 1889, and placed under the supervision of Government in 1891. In 1896 the Gārulia Municipality was formed from a portion of North Barrackpore, and took over the dispensary, which it entirely maintains. It treats in-patients and out-patients, and contains three beds. There are no private subscriptions.

A large proportion of the inhabitants of Gārulia are workers in the two large factories there, the Shāmnagar Jute Mill and the Dunbar Cotton Mill. Each of these factories keeps up a dispensary, under a competent officer, for the treatment of its hands.

Hālishahar.—Opened in 1887. Its full title is the "Bayley Jubilee Charitable Dispensary, Hālishahar." It is entirely maintained by the Municipality, and treats out-patients only. There are no subscribers. It is accommodated in a good *puccā* building, built by Lieutenant-Colonel K. P. Gupta of the Indian Medical Service (whose home was at Hālishahar), who presented it to the Naihāti Municipality, on condition that it would maintain it. There is one large room used as a compounding and consulting room, with two smaller ones used as a female waiting room and as a store room for medicines and instruments. It is almost on the bank of the Hooghly and is about two miles (by an indifferent road) from Hālishahar station, on the main line of the Eastern Bengal State Railway (26 miles from Sealdah) but can more conveniently be reached by road from the much larger station of Naihāti, four miles off, by a good *puccā* road.

Harināvi (*Rājpur*).—Opened in March 1877. It is maintained by the Rājpur municipality, with a subvention of Rs. 10 per month from the District Board. It treats out-patients only. The dispensary is held in the Municipal office and is situated about 12 miles from Calcutta by a good *pucca* road. Going on along this road, Bāruipur railway station is reached after four more miles, and Bāruipur Dispensary after another mile. The Harināvi Dispensary is also about three miles, by a good *pucca* road, from Sonārpur station on the Diamond Harbour branch of the Eastern Bengal State Railway, ten miles from Sealdah.

Jaynagar.—Opened in 1899 under the title of the Allen Charitable Dispensary, being so named after the late Sir Charles Allen, who was then District Magistrate of the 24-Parganas. It is maintained by the Jaynagar Municipality, and receives grants from the Government Khās Mahāl Fund and the District Board. There are no private subscribers. It treats out-patients only. The dispensary is accommodated in a *pucca* building belonging to the Jaynagar Municipality, which also accommodates the post-office and the municipal office. At first, the dispensary was held in a small enclosed verandah, but in 1899 a fair-sized room was allotted to it.

Jaynagar is the most inaccessible dispensary in the district. The usual way of reaching it is to go by rail from Sealdah to Magra Hāt station on the Diamond Harbour branch of the Eastern Bengal State Railway, and thence along the Ganga Nullah to Jaynagar. Even at the height of the rains this nullah, though an old bed of the Ganges, will not float any boat of larger size than a *donga*, or dug-out, made from the trunk of a tree; and from February to April it has not enough water to carry even such humble craft. The distance is seven miles, and a dug-out, with two men poling, takes from 2 to 2½ hours to cover the distance. Jaynagar is about 17 miles by *pucca* road, from Bāruipur station, but this road is only passable for wheeled traffic in the cold and hot weather. There is no direct road from Magra Hāt, but only tracks across the fields and a roundabout road, which, after 5 miles, strikes into the Bāruipur-Jaynagar Road, at Dakshin Bārāset, four miles from Jaynagar.

Kāmārhāti.—This hospital, which is known as Sāgar Dutt's Hospital, was opened in 1899. It treats both in-patients and out-patients. It is maintained by endowments left by Babu Sāgar Dutt, who left by will the following:—(1) a garden with an area of 110 *bighas*; (2) one lakh for building; and (3) property producing about Rs. 25,000 a year for the maintenance of a hospital and school, the former having the prior claim. The buildings (which are all excellent *pucca* buildings) are as follows:—(1) An out-patient department for males, with a private examination room, rooms for medicines and compounder's quarters, close by the gate leading into the grounds from the Grand Trunk Road to Barrackpore. (2) A male hospital, in the centre of the grounds, which contains two large wards, medical and surgical, with 12 beds each, 5 smaller wards, with two to four beds each, an operation room, office, compounder's duty room and store-room. (3) A female hospital, towards the north-west of the grounds, having one ward with six beds, two smaller wards with two beds each and quarters for the female hospital assistant on an upper storey. (4) A female out-patient department, close to the female hospital. These two are approached by a separate entrance from a road on the north of the grounds. (5) Assistant Surgeon's quarters, *post-mortem* room, servants' quarters, etc., not far from the male hospital. A school building was put up in 1898 in the extreme north-west corner of the grounds, with a separate entrance. The grounds lie on the west of the Grand Trunk Road to Barrackpore, 8½ miles from Government House. They are about 1½ miles (by a *pucca* road) from Belgharia station, on the Eastern Bengal State Railway main line, which is seven miles from Sealdah. The hospital is centrally situated for a large population in Kāmārhāti, Barnagore, North Dum-Dum and South Barrackpore. The medical officer is a Government Assistant Surgeon.

Kānchrāpāra.—This dispensary is maintained by the Eastern Bengal State Railway, and the Civil Surgeon of the 24-Parganas has nothing to do with it. It treats only out-patients, who are railway employés.

Khardah (South Barrackpore).—A dispensary was opened as Agarpārá (in South Barrackpore) in 1873, and was removed to Khardah, about two miles further north, in 1889. It is maintained by the South Barrackpore Municipality, and treats out-patients only. It is accommodated in a *pucca* house, with three rooms and a verandah, located in a garden which belonged to the late Māhārāja Nil Krishna Deb Bahādur. The house is not the property of the dispensary or municipality; they are only allowed the use of it by the estate. There is an upper storey, which is let (by the estate) for use as a dwelling house. The building is on the east side of the Grand Trunk Road to Barrackpore, about 12 miles from Government House. It is about one-third of a mile from Khardah station, on the Eastern Bengal State Railway main line, 11 miles from Sealdah.

The Khardah Jute Mill keeps up a dispensary, about a mile from the municipal dispensary. The Titāgārh Jute Mill, Titāgārh Paper Mills and Standard Jute Mill, all situated in Titāgārh Municipality, which is close to South Barrackpore, also maintain dispensaries for the use of their own hands. The Bhola Nāth Bose Dispensary at Barrackpore is situated within the limits of the South Barrackpore Municipality.

Kidderpore Dockyard (Royal Indian Marine Dispensary).—Opened in 1878, and entirely maintained by the Royal Indian Marine Department. It treats out-patients only, and only employés of the dockyard; hence, the patients are all male adults. It occupies the upper storey of a small two-storeyed building, on the left of the gate of the Royal Indian Marine Dockyard, about 500 yards from Hastings Bridge over Tolly's Nullah.

Kidderpore Municipal Dispensary.—Opened in 1891. It is entirely maintained by the Corporation of Calcutta, and is under the supervision of the Health Officer and Assistant Health Officer of the city, not under the Civil Surgeon of the 24-Parganas. It treats out-patients only. Until 1899 this dispensary was held in a small building, up a court of Garden Reach Circular Road, on the west side, about half a mile from the Kidderpore Bridge over Tolly's Nullah. In 1899 it was transferred to a large two-storeyed house on the opposite side of the road.

Māniktala (Nārkaldānga).—Opened in 1895, and brought under Government supervision in 1898. It is entirely maintained by the Māniktala Municipality; there are no subscribers. It treats out-patients only. The dispensary was held in the municipal office, a hired building, in Bāghmāri Road, Māniktala, up to August 1889, when the municipal offices were moved into a building, newly constructed for the purpose, on the north side of Nārkaldānga main road, about one-and-a-half miles from Sealdah. The dispensary occupies the east side of the ground floor of the building, and has separate rooms for use as a consulting room, compounding room, female waiting room and private examination room. In September 1899 the Māniktala Municipality opened a branch dispensary at Ultadānga (in the north of the town), the arrangement being that it should be opened in the afternoon only and be worked by the staff of the Māniktala dispensary, which was to be closed in the afternoon.

Naihāti.—Opened in 1874. It is entirely maintained by the Naihāti Municipality. There are no private subscribers. It treats out-patients only. It is accommodated in a *pucca* building, with thatched roof, containing two rooms. A female waiting room is provided by a second *kutcha* building, a few yards to the east. These buildings stand in the municipal office compound about ¼ mile north of Naihāti railway station, on the west of the road to Kānchrāpāra.

North Dum-Dum (*Birāti*).—Opened in 1883 at Nimta, about 1½ miles from Belgharia, but transferred about 10 years ago to Birāti, where a new building was constructed for the purpose.

Russa (*Prince Ghulām Muhammad's*).—Opened in 1875; treats both in patients and out patients, and has 22 beds. It is entirely supported by an endowment (Prince Ghulām Muhammad's fund), in the hands of the Collector of the 24-Parganas. The original gift amounted to Rs. 1,50,000, out of which Rs. 25,000 was paid for the house and grounds. With the investment of some accumulated interest, the fund now amounts to Rs. 1,37,500, yielding Rs. 5,486 a year. Whatever balance is over, after paying for the maintenance of the Russa Dispensary, goes to the Chetla Dispensary. It is managed by a committee, consisting of the Judge, Magistrate, Civil Surgeon, District Engineer and Treasury Officer of the 24-Parganas, *ex-officio*, a representative of the original donor's family and two members of the Calcutta Corporation, appointed on account of their grant to the Chetla dispensary, which is managed by the same committee. The dispensary occupies a two-storeyed *pucca* building, with large grounds, at Tollygunge, on the east of the Russa Road, about three miles south of the Calcutta Cathedral.

Sealdah Railway Dispensary.—A private dispensary existed here when the railway was first constructed, in 1859. The railway then belonged to a company; and when Government took over the line in 1884, it continued to maintain the dispensary, which treats out-patients and railway employés only It is under the Medical Officer of the Eastern Bengal State Railway, and the Civil Surgeon of the 24-Parganas has nothing to do with it.

South Dum-Dum Dispensary—Was opened in 1885, but its existence was not formally sanctioned by Government until 1898. It is entirely maintained by the South Dum-Dum Municipality and treats out-patients only. It occupies two rooms in the municipal office, a *pucca* hired building, in Nagar Bazar, South

Dum-Dum, on the west side of the Jessore Road (a good *pucca* road), about four miles from Sealdah. It is about a mile from Dum-Dum Junction, the first station on the Eastern Bengal State Railway main line, four miles from Sealdah.

Tāki (Tāra Sankar Chaudhuri's).—Opened in 1865, according to the Surgeon-General's annual report for that year. An inscription on the gate states that it was opened in 1867; probably the present building was first occupied in that year. It is accommodated in a one-storeyed house, with two rooms that are used for the dispensary, and two for the medical officer's quarters, a verandah and female waiting room. The dispensary was founded by a zamīndār of Tāki, Tāra Sankar Chaudhuri, who gave the building and a subscription of Rs. 40 per month towards its upkeep. After his death the management of the dispensary was in the hands of a committee, and his widow let her subscription, on which it relied chiefly for its maintenance, fall into arrears. The affairs of the dispensary went from bad to worse, the buildings fell into disrepair, the pay of the staff was months in arrear, and there was hardly any medicine in the dispensary. Such was the condition of affairs in 1898, when the District Board stepped in, and, with the consent of the founder's widow, took over the dispensary to save it from collapse. It has an invested capital of Rs. 500, yielding interest of Rs. 23, and the balance of its cost is met by the Municipality and District Board, and by private subscriptions amounting to Rs. 500.

Other dispensaries.—The dispensaries at Belpukur, Garden Reach, Magra Hāt, and Tentulia were established during the quinquennium ending in 1904-05, and those at Bhātpāra, Hābra and Kākdip in the quinquennium ending in 1909-10. The dispensary at Bhātpāra is for women only, and is in charge of a lady doctor. Those at Tentulia and Belpukur receive grants of Rs. 845 and Rs. 375 a year, respectively, from the Government Khās Mahāl Fund. Another recent addition is the Vishnupur Dispensary, which is called the Stevenson-Moore Charitable Dispensary after Mr. C. J. Stevenson-Moore, C.V.O., formerly District Magistrate of the 24-Parganas.

The following short notes give some account of dispensaries which have at one time or another been under Government supervision, but which, owing to transfer or closure, are so no longer.

Agarpāra.—Some eight miles north of Calcutta, in what is now the South Barrackpore Municipality. It was a branch dispensary which treated out-patients only and was removed in 1889 to Khardah, where the dispensary still exists.

Baliāghāta.—In the Baliāghāta Road, in Māniktala, about a mile east of Sealdah, was opened, without sanction, by the Māniktala Municipality, in 1885. When the Māniktala Dispensary was brought under Government supervision in 1898, that at Baliāghāta was closed.

Bhawānipur.—See the account of Sambhu Nāth Pandit Hospital. It treated out-patients only, and was closed when Government aid was withdrawn in 1882.

Deganga.—At the head-quarters of the Deganga thana, about 12 miles from Bārāset, on the road to Basirhāt. It treated out-patients only. It was closed in 1869; the report on charitable dispensaries for that year says that the purpose for which it was opened had been fulfilled.

Dhānkuria (*Shāma Sundari's*)—Was opened in 1888, and placed under Government supervision in 1892. It was maintained by Musamāt Shāma Sundari Dasi, mother of Babu Upendra Nāth Sahu, banker and zamīndār of Dhānkuria. This lady died in 1899, after which her son carried on the dispensary. Dhānkuria is situated to the south-west, and just outside the limits, of the Bāduria Municipality.

Garden Reach (*Arratoon Apcar's*).—Established in 1871 by Mr. Thomas Apcar, a wealthy merchant of Calcutta. It treated out-patients only, and was closed in 1880, when Government aid was withdrawn. The building, with inscription, may still be seen in the compound of No. 21, Garden Reach Road, now the Trinidad and Fiji Emigration Depôt.

Gobardānga.—At the town of that name, a station on the Eastern Bengal State Railway, in the north-east of the district, 36 miles from Sealdah. It was founded in December 1860 by the Mukharji family, zamīndārs of Gobardānga, and was withdrawn from Government supervision in 1889.

Kulpi.—A village on the Hooghly, situated eight miles south of Diamond Harbour. An unqualified Indian doctor settled at Kulpi about 1880 and opened a druggist's shop. There was then a large estate at Kulpi, known as the Bhawānipur Ward's Estate, under the Court of Wards. This estate paid the doctor Rs. 5 per month to treat raiyats of the estate free of charge and give them medicines at half price. When Government issued orders, in 1894, that dispensaries wholly maintained by estates under the Court of Wards should be placed under supervision, this so-called dispensary was placed on the Government list, and its statistics (about one patient per day) were published among those of charitable dispensaries. It was inspected for the first time in 1898 by the Civil Surgeon, who reported that no charitable dispensary existed at Kulpi, and recommended that it should be struck off the list. This was done in 1899, when the Bhowānipur ward came of age, and the management of his estate was made over to him by the Court of Wards.

Rājārhāt.—In Bhāngar thana. It treated out-patients only and was closed in 1869 for want of funds.

Vishnupur.—At the head-quarters of the Vishnupur thana, on the Diamond Harbour Road, 18 miles south of Government House It was opened in 1869. The report on charitable dispensaries for that years says it "took the place of Rājārhāt."

There were 63 licensed vaccinators in 1911-12, when 62,522 primary vaccinations were performed, of which 61,982 or 99 per cent. were successful. The average annual number of persons successfully vaccinated in the previous five years was 61,531 or 32·6 per mille of the population. There is very little opposition to vaccination, but there are difficulties in inspecting the work of the vaccinators owing to the distance and inaccessibility of some of the places in the interior. VACCINATION.

The Civil Surgeon is *ex-officio* Medical Inspector of emigrants to the Colonies, and a Sub-Assistant Surgeon is Assistant Superintendent of Emigration (inland) at Naihāti. A special officer is certifying surgeon for factories in the Barrackpore subdivision, with the exception of Dum-Dum thana, where the duties of that officer are performed by the Station Staff Surgeon. The Civil Surgeon of the 24-Parganas is certifying Surgeon for the rest of the district, *e.g.*, for the Budge-Budge jute mills, the Empress of India cotton mill, the bulk-oil depôt at Budge-Budge, the Lower and Upper Hooghly jute mills, the Garden Reach cotton mills, the Clive jute mills and the oil factories of Messrs. Graham & Co., and Messrs Shaw, Wallace & Co at Budge-Budge. The Civil Surgeon is further *ex-officio* Consulting Physician to the Alipore Central Jail, and also for the purposes of Government life insurance business carried on by the postal department. MISCELLANEOUS.

CHAPTER V.

AGRICULTURE.

GENERAL CONDITIONS. THE arable land in the 24-Parganas may be divided into three main classes, viz., (1) the comparatively high land along the banks of the rivers, (2) the low-lying depressions that stretch away below the river banks and (3) the lands adjoining the Sundarbans, which have been reclaimed in recent times. In the riparian tracts first mentioned the land is fairly well raised, and the drainage passes away easily to the basins below. It has been elevated by the deposit of silt from the rivers in past generations, and has attained a height which ensures it against inundations, but at the same time prevents it from receiving the fertilizing layer that the floods formerly left behind them. The low lands that constitute the greater portion of the cultivated area are mainly under rice and jute. Large areas are occupied by *bils*, which may be either large fresh-water lakes or waterlogged swamps. Some are natural drainage basins that never dry up, and cannot be drained owing to their level. Others are connected with the rivers by efficient water channels, which serve two useful purposes; they bring down rich river silt, and they drain away the surplus water from the *bil*. In many cases, however, the creeks or *khāls* have been silted up and have ceased to perform these functions. The cultivators are, therefore, forced to wait till the *bil* lands dry up, and, if there is a year of heavy rainfall, are precluded altogether from attempting their tillage. In the third tract, *i.e.*, in the reclaimed lands adjoining the Sundarbans, cultivation is only rendered possible by means of embankments constructed to keep out the salt water. Dams (*bāndhs*) have also to be built across the *khāls*, so as to prevent the ingress of salt water from the rivers with which they communicate. The soil in this last tract being impregnated with salt, fairly heavy rainfall is necessary to wash it out before rice seedlings can be grown and transplanted.

RAINFALL AND IRRIGATION. In the district as a whole, cultivation suffers far more frequently from excessive, than from deficient, rainfall, for, with the exception of strips of high land along the banks of the rivers, the country is low and swampy, and tends to become waterlogged whenever there is heavy rainfall. This is especially the case with the great basin shut in between the Diamond Harbour Railway and the Hooghly embankments, as well as a similar tract east of the Eastern Bengal Railway and the Balli Bil : in these and other cases the natural drainage channels are inadequate to remove the volume of water which accumulates after heavy precipitation of rain. It will readily be understood that, in these circumstances, there is little necessity for artificial irrigation. It is, in fact, only resorted to for the cultivation of sugarcane and garden crops, for which water is raised from tank and ditches.

The soils of the district belong to four main classes, viz., *mātiāl* or clayey soil, *dorasa* or loamy soil, *bālia* or sandy soil and *nona* or saline soil. *Mātiāl* is further subdivided into three varieties called *kāla mātāl*, *rānga mātāl* and *jhājhra mātāl*. *Kāla mātāl* is a stiff black clay of great natural fertility, on which all kinds of crops can be grown. *Rānga mātāl* is of a reddish colour: it cracks in the dry season and sinks into holes in the rains. It is well suited for winter rice, and on higher levels can be used for the cultivation of jute and other *bhadoi* crops. *Jhājhra mātāl*, which is inferior to the other two varieties, is blackish in colour and is easy to plough even when dry. SOILS.

Dorasa soil is a mixture of clay and sand. It is used for *bhadoi* and *rabi* crops and also suitable for sugarcane. The *dihi* lands, or elevated lands surrounding village sites, come under this category. Being generally highly manured, they are devoted to sugarcane, tobacco, red pepper and vegetables. *Bālia* is a common name for all soils in which the proportion of sand exceeds that of clay. Such soils are used for tobacco, potato, *āus* rice and *mung* (*Phaseolus mungo*). *Nona* is a wet saline soil, which in ordinary years does not dry up enough to permit of cultivation. It is only when the rains are late that it dries up sufficiently for cultivation to be possible.

The soils in the Sundarbans, where winter rice is practicably the only crop, may be divided into the following four classes. *Mātiāl*, a clayey soil, whitish in colour, and loose and light in composition. This soil is very suitable for the "Patna" rice which is grown so largely in the Sundarbans. Next in quality comes a loamy soil called *bāliāra* or *dorasa*. It is reddish in colour and will retain moisture longer than any other soil. Coarse paddy is grown on it, but not very profitably. *Dhāpa* or *chura* is a soil of a whitish colour, which lies at higher levels than the other classes. Consequently, it is not covered with water, and the salt is not washed out, unless there is heavy rain. In ordinary years therefore no crops can be grown on it, and it only bears *ulu* grass, which is used for thatching. Paddy can be grown on it when the salt is washed out by heavy rain, but the yield is usually small. *Dhāl* is the lowest land of all and is consequently flooded earlier than the others. Like *ranga mātāl*, it is reddish in colour, cracks when dry, and is full of holes in the rains. If there is moderate or scanty rainfall, coarse paddy can be raised on it with profit, but if the rain is early and heavy, it is impossible to bring it under cultivation.

SUNDARBANS RECLAMATION AND CULTIVATION.

Cultivation is spreading rapidly in the Sundarbans, and in other parts of the district swamps are being gradually drained and reclaimed. The system of reclamation in the Sundarbans has several peculiar features, which have been well described by Sir James Westland in his Report on the District of Jessore. After explaining how the forest tress have to be cleared away and the thick brushwood hacked down, he goes on to say:—

"Unless the greatest care is taken of the land so cleared it will spring back into jungle and become as bad as ever. So great is the evil fertility of the soil, that reclaimed land neglected for a single year will present to the next year's cultivator a forest of reeds (*nal*). He may cut it and burn it down, but it will spring up again almost as thick as ever; and it takes about three eradications to expel this reed when once it has grown. The soil, too, must be cultivated for ten or twelve years before it loses this tendency to cover itself with reed jungle. When a sufficient number of people are gathered on a new clearing, they tend, of course, to form a settlement, and to remain permanently where they are. But the furthest advanced parts of the cultivation, and some also of those which are not new or remote from old lands, are carried on upon a different principle. A large number of husbandmen, who live and cultivate lands in the regularly settled districts to the north, have also lands in the Sundarbans, which they hold under different landlords.

Seasons of cultivation.

"The cultivating seasons in the Sundarbans are later than those further north. The plan which is followed by these double cultivators is as follows: The months of Chaitra, Baisākh and Jyaistha, corresponding roughly to the English months of April, May and June, are spent in cultivation at home. The husbandman then, having prepared his home cultivation, embarks with his ploughs, oxen, and food and proceeds to his *ābād* or Sundarban clearing. July, August and September are spent in ploughing, sowing and preparing the crops there, the peasant building a little shed as a dwelling for himself. The water gets high in August and September, but this is little impediment to cultivation. A considerable portion of the land under rice is situated below high-water mark; but the planting is easy, for rice shown on higher lands is transplanted into these low lands when it is strong enough to bear the waters. After having sown and transplanted his Sundarban crop, the husbandman returns home, and these outposts of civilization are absolutely abandoned —large extends of cultivated rice field without a trace of human habitation. By the middle of December, the home-cultivated rice has been cut and stored, and the peasant then returns to the Sundarbans, and reaps the crop on his clearing there. At this time of the year (January and February), reapers, or *dāwāls*, crowd to the Sundarbans and are extensively employed for the harvesting. When the rice is cut and prepared for sale, the *bepāris*, or dealers, come round and buy it up, and the zamīndār also sends his agents round to collect the rents from the cultivators. The peasant, having sold his grain, pays his rent, and brings the balance of his money back with him to his home.

"While a great deal of cultivation in the more remote parts of the Sundarbans follows this method, in the nearer tracts there are large settlements of husbandmen who dwell permanently near the land they have under cultivation. But it must be remembered that these tracts are, after all, sparsely inhabited, and that many of the cultivators who dwell in them, besides having a holding near their own houses, have also another, eight or ten miles away, which they visit only occasionally when they have work to do. The great fertility of the land renders it easy for a husbandman to keep large areas under cultivation; and thus, what with resident large cultivating husbandmen and non-resident husbandmen, the population in the Sundarban tracts is not at all equal to what the amount of land under cultivation would lead one to expect

Embankments.

"Another feature in the reclamation and cultivation of these Sundarban lands is the embankment of water inlets. It is a characteristic of deltaic formations that the banks of the rivers are higher than the lands further removed from them; and the whole of the Sundarbans may be looked on as an aggregation of basins, where the higher level of the sides prevents the water coming in to overflow the interior. Many of these basins are so formed, that, left to themselves, they would remain under flood, as they communicate with the surrounding channels by means of *khāls*, or small water-courses, which penetrate the bank; and a great part of reclamation work consists in keeping out the water, and thus bringing under cultivation the marsh land inside.

"In employing this method, all the inlets from the surrounding channels are embanked, and smaller channels, called *poyāns*, are opened round their ends. The inlets themselves are too big to be kept under control, but these *poyāns* can easily be so kept. This embanking is usually done in November, after the rivers have gone down. When the tide is low, the channels are opened, and the water from the inside drains off; when it is high, the channels are closed. Much land can be rendered culturable by this means, which would otherwise be marsh. But here also a single year's neglect may take away at one stroke all that has been gained by many years' labour. The effect of the rains and the freshets of each year is to partially destroy all the embankments that were used the previous year and to flood the lands. The rice that has been sown has, however, attained sufficient hardihood to remain uninjured; and when the waters again go down, the harvest may be reaped. But unless the embankments are again renewed in November, the floods will not have ceased to cover the low lands by sowing time, the land will remain unsown, and jungle and marshy reed will take the place of the paddy."

AREA UNDER CULTIVATION. Owing to the large extent of waste land included in the Sundarbans, where the forests under the administration of the Forest Department alone cover 1,711 square miles, the proportion of cultivable land is small, being, in fact, less than half the district area. According to the returns for 1911-12, the net cultivated area amounts to 1,530 square miles, which represents 32 per cent. of the whole district, and 68 per cent. of the cultivable area. Current fallow occupies 183 square miles, cultivable waste (other than fallow) 527 square miles, and uncultivable waste (outside the forest area) 894 square miles.

PRINCIPAL CROPS. The cultivation of rice predominates, almost to the exclusion of other crops, for it accounts for 88 per cent. of the cultivated area, and other food crops for only 4 per cent. Next to rice, the most important crop is jute, which, in 1911, was raised on 133 square miles. Although, however, its cultivation has expanded greatly in recent years, the land devoted to its growth is only 8½ per cent. of the cultivated area, or one-tenth of that under rice.

Rice. The normal area under rice is 1,526 square miles, and about seven-eighths of this is *āman* or winter rice.

Aman. *Aman* rice is cultivated on low land, where water lies from one foot to three feet deep in the rains. The preparation of the land begins in the latter half of February or the beginning of March, the land being ploughed several times before sowing. In April or May, after the first fall of rain, seed is scattered broadcast in a nursery. When the seedlings make their appearance, another field is prepared for transplating. After the rainy season has thoroughly set in, the field is repeatedly ploughed until the water becomes worked into the soil, and the whole is reduced to thick mud. The young rice is then taken from the nursery, and transplanted in rows about nine inches apart. The crop is generally ready for harvesting in November or December.

Aman rice is also occasionally sown broadcast in marshy lands, *e.g.*, in parts of the Sundarbans where land suitable for nurseries is not available. Sowing takes place in the early part of July, and the crop is ready for reaping in January, the soil easily retaining up till that time all the moisture necessary for the growth of the grain. When the crop is grown in deep water it is reaped by cutting off the heads, and the straw is subsequently burnt down when the land dries up.

The finest outturn of winter rice is obtained from the reclaimed portions of the Sundarbans, which are famous for the teeming harvests obtained from the rich virgin soil.

Aus rice is generally sown on high ground. The field is ploughed when the early rains set in, ten or twelve times over, till the soil is nearly reduced to dust, and the seed is sown broadcast in April or May. As soon as the young plants reach six inches in height, the land is harrowed in order to thin the crop and clear it of weeds. The crop is harvested in August or September, and a second crop of pulse or oil-seeds is generally taken off the land in the cold weather. ***Aus.***

Boro rice is a comparatively unimportant variety sown on marshes which dry up in winter. The preparation of the land commences in the middle of November; sowing takes place ten days later; and reaping lasts from the middle of March till the middle of April. The land is hardly ploughed at all. The seed is scattered broadcast in the marshes or *bils* as they dry up, and the young shoots are transplanted when about a month old. There is another kind of *boro* paddy called *āus boro*, which is sown broadcast during the months of April and May, and reaped in August or September. ***Boro.***

Another description of rice, known as *uri dhān*, is indigenous in the deep-water marshes, and is occasionally used as food by fishermen and boatmen. The plant looks like a confused mass of creepers floating on the water, and shoots forth its ears of grain in every direction. A peculiarity of this rice is that the grain drops from the ear into the water when it attains maturity. To prevent this, the ears are bound together before ***Uri dhān.***

the paddy ripens. This rice grows plentifully in the marshes, but very little is collected, for the swamps are deep, and the crop hardly repays the labour of binding the ears and collecting the grain.

Other cereals and pulses.

Except rice, there is no cereal of any great importance. Gram (*chholā*) is cultivated on high land, but in 1911-12 had an area of only 500 acres, while other cereals and pulses were grown on 29,900 acres. They are mainly cold weather crops, such as peas, *masuri*, *khesāri* and *kalāi*, sown in October and gathered in February or March, and are cultivated only on small patches of land.

Jute.

The extensive cultivation of jute dates back only half a century. Its introduction is described as follows in the Statistical Account of Bengal, Volume I, published in 1875: "Jute was formerly very little cultivated in the district, and the small quantities that were produced only sufficed for local requirements, such as rope-making, etc. About fifteen years ago a great demand arose for gunny cloth, and almost every family that could get a little money to establish a jute loom did so. Nearly all the day-labourers took to weaving, and the women and children to spinning the thread, causing the rates of labour to suddenly rise. Jute cultivation accordingly received an impetus; and although the gunny-weaving was carried to excess, and ruined many families in the 24-Parganas, other outlets for the jute fibre sprang up, and its production has steadily increased. A large part of the high lands in the district, formerly devoted to *āus* rice, is now entirely given up to jute cultivation."

The actual quantity of land given up to the crop varies considerably according to the prices which the produce commands in the market, but on the whole it shows a steady tendency to increase. In 1901-02 the normal acreage under jute was only 63,400 acres, the actual area under the crop in that year being 56,000 acres, whereas the corresponding figures for 1911-12 were 76,600 and 85,000 acres respectively.

The seasons for sowing and growth are the same as for *āus* or early rice. After the usual ploughing, the seed is sown broadcast from the middle or end of March to the beginning of June, and the plant is generally cut from the middle of August to the middle of October, by which time it has attained a height of five to ten feet. The stalks, when cut, are made up into bundles and immersed in some pool, tank or stream, and left to steep; this process is called retting. While the bundles are under water, they are examined from time to time to see how far decomposition has proceeded. As soon as it is found that the fibre will peel off easily from the stem, the bundles are taken out, and the stalks are beaten or shaken in the water till the glutinous substance in the bark is entirely washed away. The fibre is then dried in the sun, and, when dry, is made up into hanks (*gānt*) and sold to agents, who consign it to the jute presses and mills.

Sugarcane and date palms. Sugarcane occupies a considerable area in the north-east of the district, where also the cultivation of sugar-yielding date palms (*khejur*) is carried on extensively. These trees are planted in regular rows, the plantations being generally laid out on land which is too high for the successful cultivation of rice. The regulation distance between the trees is about 12 feet, so that the number in a plantation of an acre will be about 300. The tree is tapped when it is "ripe," *i.e.*, when it is seven or eight years old, and is carried on in the cold weather. When the rainy season is over, and there is no more fear of rain, the cultivator cuts off the leaves growing out of the trunk for one half of its circumference, and thus leaves bare a surface measuring about 10 or 12 inches each way. This surface is at first a brilliant white, but becomes by exposure quite brown, and has the appearance of coarse matting. The leaves are cut off by a man who climbs up the tree supporting himself by a strong rope, which he passes round the tree and his loins. He slides the rope up and down with his hands, setting his feet firmly against the tree, and throwing the weight of his body on the rope. In this manner, his hands are free, and he cuts the tree with a sharp knife like a billhook.

After a few days, the "tapping" is performed by making a cut in the exposed surface, in the shape of a broad V, and then cutting down the surface inside the angle thus formed. The sap exudes from this triangular surface, and runs down to the angle, where a thin bamboo is inserted in order to catch the sap as it drips down and carry it out, as by a spout. Below the end of the bamboo an earthenware pot is hung at sunset, and the juice of the tree runs down into it during the night. The pots are taken down in the morning, before sunrise, as the heat of the sun closes the pores of the wood and prevents exudation during the day. The juice is extracted three days in succession and then the tree is allowed to rest for three more days, after which the juice is again extracted.

The next process consists of boiling the juice, and this most ryots do for themselves, usually within the limits of the palm grove. Without boiling, the juice speedily ferments and becomes useless; but when once boiled down, it may be kept for long periods. The juice is therefore boiled at once in large pots placed on a perforated dome, beneath which a strong fire is kept burning, the pared leaves of the trees being used with other fuel. The juice, which was at first brilliant and limpid, now becomes a dark brown half-viscid, half-solid, mass called *gur*, which is easily poured, when it is still warm, from the boiling pan into the earthenware pots in which it is ordinarily kept. It is then sold to refiners, and manufactured into sugar.

Tapping continues year after year, alternate sides of the palm trees being used in alternate seasons. Each season's cutting is above that of the previous season (but on the opposite side of the trunk), so that the trunk has a curious zigzag appearance. The age of a tree can at once be ascertained by counting the notches and adding six or seven for the number of years that pass before the tree is first tapped. Some trees have over 40 notches, showing that they have been tapped for as many years, but these are exceptional.

Tobacco.

Tobacco is generally grown for domestic use only, but in the north of the Bārāset subdivision it is largely grown for trade and export. Here it thrives on old indigo lands and may be seen planted up to the very edge of the ruined vats. The variety most grown is called Hingli, from a village of that name on the left bank of the Jamuna river. The leaf is said to sell for Rs. 5 to Rs. 7 or Rs. 8 per maund, but some special qualities fetch as much as Rs. 20 a maund.

Light soil is usually selected for the plant, and is well ploughed and manured. The seed is sown in nurseries in August and transplanted early in October, the plants being placed about 18 inches apart. They are topped when they have 12 or 13 leaves, and then suckering goes on regularly until January, when the plants are ready for cutting. This stage is reached when the leaves hang down, turn colour and have spots on them. They are cut up into pieces, each with two to five leaves on it, which are spread out in the sun to dry for a few days. They are then hung up on grass ropes or strings in the house or cowshed, and left for two months, until the south wind sets in and the leaves are 'in case.'

The next process is sweating or fermentation, which is effected in the following way. The tobacco is piled up in heaps covered with gunny sheets and resting on a layer of straw; the object of the latter is to protect them from damp. Fermentation sets in, and when the proper temperature is reached, the heap is broken up and rearranged to prevent overheating, *i.e.*, the top and bottom leaves are placed in the centre, and the inside leaves are put on the outside, so that fermentation may proceed evenly.

The cultivation of garden crops, and especially of *pān* and *baigun*, is of considerable importance in the 24-Parganas, from which Calcutta obtains a large portion of its supplies. There are numerous large market gardens in the neighbourhood of that city, where vegetables of all kinds are grown; and in many places trenching grounds are employed for the purpose with excellent results. **GARDEN CROPS.**

The betel leaf creeper called *pān* is grown in gardens, known as *baraj*, mainly by members of the Bārui caste, with whom its cultivation is an hereditary occupation. The garden is laid out on high land in the vicinity of a stream or tank. It is enclosed by a wall of bamboo and reed work, about five or six feet in height, and covered over with the same material, the roof being supported by uprights from within. This is done to protect the creeper from the sun, and to prevent cows, goats, etc., from destroying it. The enclosure thus prepared is divided off into parallel ridges about eighteen inches apart, in which are placed, about a span apart, uprights of thin bamboo, and across these a framework of other thin bamboos. *Pān.*

The land requires to be well dug up previous to planting. Cuttings are planted in February and March, and in four months, *i.e.*, by June or July, have grown sufficiently for the leaves to be plucked. During the hot months of April and May, the plants must be watered morning and evening, and in June and July the land is well manured. Fresh earth has to be put round the roots at the same time. There are three crops during the twelve months, which are called by the names of the months in which they are plucked, viz., Kārtik *pān*, Phālgun *pān* and Ashār *pān*. The first is the best; the last is the heaviest crop, but inferior in quality. When plucking, it is a rule always to leave sixteen leaves on the creeper. The leaves are eaten in the green state with betel-nut (*supāri*), lime and cardamoms: the first green leaves, especially those plucked in the early spring, are preferred. The cultivation of the plant requires constant care, but is highly remunerative. A garden lasts 5 to 20 years.

The brinjal or egg-plant (*baigun*) is cultivated extensively, and is of considerable economic importance. The two main varieties are called *āus* and *paus*, but each may be subdivided into a number of species, such as *muktakeshi* and *elakeshi*, according to size and colour. A third variety, which is not so commonly grown, is known as *duko* or *kuli baigun*; this is much smaller in size and grows in bunches. The plant grows in sandy loam and also in clay soils, but the land must be high and well **Brinjals.**

drained. It grows best in fields that have been left fallow for a year or two.

For the cultivation of *āus baigun* the land, if not a fallow, must be well manured, e.g., with mud taken from the beds of tanks, with which cow-dung may be mixed. Several ploughings are necessary, and the soil has to be levelled, weeded and pulverized. The seedlings are transplanted at the end of Chaitra or Baisākh, and placed about 3 feet apart. They take root in about 10 days, after which the space between the rows is hoed, and the plants are earthed up, the furrows being converted into ridges 3 or 4 inches high. After another fortnight, the field is weeded and the plants are again earthed up, so as to make the ridges 9 inches high. Three weeks later, after another weeding, the plants are earthed up for a third time to a height of 15 or 18 inches. They begin to bear fruit at the end of Ashār or the beginning of Srāban, and continue to do so till Phālgun.

The cultivation of *paus baigun* follows a different method. It grows best on a clay soil, and thrives in the Bāruipur thana, the brinjals of which have a special repute. The seeds are sown in a nursery in Jyaistha or Ashār, the fields in which they are to be transplanted being prepared in the same way as for *āus baigun*. Transplantation is carried out in Srāban, the seedings being placed in rows 3 feet apart. The land is hoed two or three times, and is weeded as occasion requires. The plants begin to bear fruit in Agrahāyan, and continue to do so till Chaitra.

Kuli baigun is sown in Aswin and Kārtik, and planted in Agrahāyan and Paus.

Other vegetables.

Four varieties of the *kachu* (yam) are grown as field crops, viz., *man*, *mangiri*, *kachurmuki* and *solakachu*, and two varieties of *patal*, viz., *paikhati* (or *chakdayi*) and *deshi*. The gourd called *kumra*, of which there are two kinds, viz., *deshi* and *belāti*, is very common; its creepers may be found in nearly every house, either climbing on the thatched roof or trailing on bamboo stages made for the purpose. The following vegetables may also be seen in the ryot's vegetable garden, which is always in or near his homestead:—radishes (*āus* and *paus mula*), *uchchi* (*karala* and *uchchi*), *jhinge* (*teta* and *pala*), *tarmuz* (*deshi* and *teta*), sweet potatoes, spinach, cabbages, cauliflowers, cucumbers, onions and garlic.

FRUITS.

Cocoanut trees grow abundantly, especially in the south of the district; the fruit is collected in the rainy season. A fully ripe cocoanut is called a *jhunā*, and from its kernel several kinds of sweetmeats are made, such as *nārikel nāru*, *raskarā*, *chandrapuli*, etc. The nut is put to a variety of uses; ropes and mats are made from the husk; oil is extracted from the kernel; the shell is made into the bowls of *hookahs*, cups, etc.; and the tree itself, when past bearing, can be cut down, and the trunk hollowed into a canoe. Plantain trees are grown extensively: the variety known as *champa* has a deservedly high reputation for the delicious flavour of its fruit. Of other cultivated fruits the following may be mentioned: the mango, papaya, jack, guava, custard-apple, plum, *bel*, tamarind and pine-apple.

CATTLE. The cattle belong to the degenerate breeds common in Lower Bengal. They are said to be deteriorating owing to the cultivation of pasture lands and to the abandonment of the practice of dedicating bulls. Diminutive goats are numerous, but ponies, sheep and buffaloes are scarce. There is a Veterinary College at Belgāchia; and the Mārwāris maintain a *Pinjrāpol* or asylum for broken-down cattle at Sodepur, 10 miles north of Calcutta, where they have an annual gathering in November.

CHAPTER VI.

DRAINAGE AND EMBANKMENTS.

DRAINAGE WORKS. The drainage of low-lying areas is a question of considerable importance in the 24-Parganas, particularly in the country to the south of Tolly's Nullah, where large tracts are so swampy and water-logged that artificial drainage is necessary to make them cultivable. Much has already been done for the drainage of land in this neighbourhood, 30 basins, as shown below, having been effectively drained :—

Basin.		Area (in square miles).	Basin.		Area (in square miles).
Ahmadpur	...	5	Pānchpota	...	3
Birāsi ...	...	3	Pātnighāta	...	4
Burul ...	...	5	Rāypur ...	...	2
Chāriāl ...	...	37	Sadiāl ...	...	4
Chitāmāri	...	1½	Sāgar ...	...	½
Dhosā ...	...	1	Sāmukpota	...	½
Habkā ...	...	50	Sātpukur	...	43
Hārā ...	...	5	Srīkrishnapur	...	3½
Haral ...	...	½	Surjipur	...	25
Kalābāria	...	1¾	Tengrābāria	...	5½
Kulpi ..	...	28	Tengrābichi	...	17
Kāorāpukur	...	2½	Telāri ...	...	5
Kholākhāli	...	11	Tetulia ...	...	3½
Magrā Hāt	...	219	Tilpi ...	...	1
Nishāntala	...	5	Uttarbhāg	...	5

The most important of these works are those designed for the drainage of the Magrā Hāt, Habkā and Surjipur basins, of which a detailed account is given below. Of the others the following may be mentioned :—

(1) The Chariāl works, carried out in 1887, drain a large area (the Chariāl Bil) in the neighbourhood of Budge-Budge.

(2) The Bâlli Bil in the north-east of the district is drained by a channel with a sluice at Tentulia, the work being completed in 1882.

(3) The Sātpukur, Kulpi and Tengrābichi works drain an extensive area in the south of the Diamond Harbour subdivision.

MAGRA HAT DRAINAGE SCHEME.

By far the most important drainage scheme as yet instituted in the district is the Magrā Hāt scheme, which provides for the drainage of nearly 300 square miles of country in the vicinity of Diamond Harbour and Magrā Hāt by means of a system of channels, with controlling sluices, which are designed to serve both as drainage and navigation channels. The country benefited by the scheme forms part of a large area, which is surrounded by a continuous embankment, known as the 24-Parganas embankment and in part as the Hooghly left embankment. Starting just below Akrā on the left bank of the Hooghly river, it forms the left flanking embankment of that river for some 78 miles, as far south as Chitāmāri, where it turns towards the east and then to the north-east, forming a protection to the country from the large tidal creeks of the Sundarbans. A good deal of reclamation has been carried out here, and the main embankment is to a large extent marked by new bunds to the south of it. Reaching the Piāli river, the embankment turns northwards to the confluence of the Bidyādharī river, which it flanks, then bends westwards along Tolly's Nullah and ends at Gariā, some 11 miles east of its starting point, after a course of about 212 miles. The whole enclosed area is about 717 square miles, of which the tract affected by the Magrā Hāt scheme occupies mainly the eastern and central portions. A striking feature of the enclosed tract is its uniformity of level. There is no general "trend" of the ground in any direction, except locally, so that the drainage as a rule follows the direction of what are, for the time being, the most efficient outfalls. The embankment is pierced by numerous open *khāls*, which serve as exit-channels for rain-water. Until the execution of the drainage scheme, only some of these channels had sluices; in the other channels protection was afforded by "returning" the embankment along their flanks. The most important of the sluices in question are as follows:—

The 5-vent Chāriāl Khāl sluice at Budge-Budge drains an area of some 37 square miles in the north-west. From there down the west side the drainage was served by open *khāls* and small sluices (mostly built for irrigation purposes) as far as the 2-vent sluice at Bendāl, built to drain an area of 11 square miles into the Diamond Harbour Creek, but really carrying drainage from a larger area. The three sluices at Kulpi, Tengrābichi and Sātpukur, with an aggregate of 13 vents, were constructed to drain a combined area of 88 square miles in the southern tract; but as a matter of fact they received drainage from a far larger area owing to a network of small *khāls*. It is probable that, since the outfalls silted, a large portion of the area which is now served by the Magrā Hāt scheme found its way to these sluices. On the south-east there are the old sluices at Khāri and Pātnighāta, and on the east the important 5-vent Surjipur sluice. The Arāpanch sluice in the north-east, serving an area of 20 square miles, completes the list of the important sluices. These sluices, it should be noted, do not provide for the drainage of the central portion of the district, which contains a large area of swamp, besides cultivated land which is periodically flooded.

Starting at the north, the main waterway is the Kāorāpukur Khāl, which takes off from Tolly's Nullah near Tollygunge: for 5 or 6 miles this *khāl* is tidal, and the drainage of this portion has to be led northwards. After this the *khāl* runs due south for some 10 miles to the village of Nainān, which is situated about the centre of a very extensive swamp. Here the *khāl* divides, the smaller branch taking a course westwards until, joining with the Srīchandra Khāl from the north, it falls into the Diamond Harbour Creek at the village of Usti. This portion of the *khāl*, which bears different names in the different villages it passes through, is now much silted, and in places hardly exists. The land along it is low in places, and there is a great swamp between Nainān and Chagdah Hāt, and another smaller swamp lying more to the west. From Nainān the main branch of the Kāorāpukur Khāl runs through swamps south-eastwards into Magrā Hāt. The Hotār Khāl joins a few miles north of Nainān, while Magrā Hāt is practically the centre of a branching system of *khāls*. Eastwards the swamps extend to Surjipur, and southwards again we come to the great Jaynagar swamp (surrounding the Jaynagar and Kātā Khāls): this is the lowest portion of the basin. South-westwards from Magrā Hāt lies the tract whose original drainage artery was the Sangrāmpur Khāl, much of which is a swamp. These are the main central swamps.

In addition, there is a tract drained by the Srīchandra Khāl, north of the Diamond Harbour Creek; while to the east of the Jaynagar swamp, across the Jaynagar *pucca* road, lies a tract of some 50 square miles, which has been included in the project under the name of the Habkā Section. The lowest part of this area is at the east corner, near Dhosā, where there are some swamps which, over a small area, are very deep and permanent, and, at a rather higher level, fairly extensive. The water, in fact, collects in a sort of pocket at this south-eastern corner of the district. This tract is intimately connected with the great Jaynagar swamp by *khāls*.

As regards the drainage outfalls, the natural, and only practicable, main outlet for the Habkā section is into the Piāli river, near Dhosā. The Srīchandra Khāl discharges into the Diamond Harbour Creek, and the natural outfall of the Sangrāmpur Khāl is into the Nāzrā Khāl, which is simply a continuation of the same creek. The country between Magrā Hāt and Surjipur, together with the low-lying tract on the east of the Kāorāpukur Khāl as far north as Hotar, was formerly served by the Surjipur Khāl, discharging into the Piāli river. About 40 years ago this *khāl* was sluiced, but the sluice was located at Surjipur, some 7 miles from the mouth of the *khāl*, with the inevitable result that the *khāl* silted up and ceased to be an efficient outlet. A considerable quantity of drainage must have gone towards this sluice, as there is continuous water-communication with the Jaynagar swamp, as well as from the Kāorāpukur Khāl. There is, however, an alternative exit from the Kāorāpukur Khāl, which is connected continuously, through Magrā Hāt, with the Sangrāmpur Khāl, which used to discharge into the creek; as well as the exit through the old *khāl* on the north of the railway from Nainān to Usti.

Thus it will be seen that all the tracts drained by the Kāorāpukur Khāl, as well as the Jaynagar swamps, had their choice of exits *viâ* Surjipur eastwards, or westwards into the Diamond Harbour Creek. Considering the number and (original) size of the channels communicating with the creek, there can be no doubt that it formed the outlet for by far the greater volume of drainage. Partly for this reason, but mainly because the permanency of the Hooghly as a drainage channel is assured, while the rivers towards the east show a marked tendency to silt up, it was decided, in the present scheme, to lead the drainage into the Diamond Harbour Creek, and to utilize the Surjipur outfall as a subsidiary exit, to relieve the main sluice in times of pressure.

The drainage scheme as now developed is the inevitable outcome of the silting up of the creeks which used to serve as drainage channels. The necessity of an efficient system of drainage had been apparent for a long time. In 1879 Mr. Whitfield, Executive Engineer, in a general report on the defective drainage of the 24-Parganas district, wrote as follows:—

"Besides the permanent *jheels*, there is, in the central part of that portion of the district around which public embankments are constructed, a large area little better than a permanent *jheel* comprising a tract of country extending over an area of about 20 square miles, having Bānkipore or Magrā Hāt as its centre. Into this area the tide flows from Diamond Harbour up the Diamond Harbour Creek and Usti Khāl, from Tolly's Nullah up the Kāorāpukur Khāl, and from Budge-Budge up the Chāriāl Khāl. In the dry weather the tides overspread the low land with salt or brackish water, and the rains inundate and destroy the crop. From want of drainage and protection, the productiveness of the locality is only a fraction of what it should be, and the inhabitants, although they may be supposed to be innured to their semi-amphibious condition by a long course of preparation resulting in the survival of the fittest, are affected similarly to those living in the vicinity of the permanent *bils*. Fever is constantly present in every village, and other classes of sickness find a congenial home in the unwholesome atmosphere prevailing in this extensive locality."

The present scheme owes its inception to an extraordinary fall of rain which occurred in September 1900. On that occasion nearly 17 inches of rain fell in one day at Diamond Harbour, and 12 inches at Surjipur; while during the week from the 18th to the 25th September these gauges recorded no less than 38¾ and 35¼ inches respectively. Even with clear drainage outfalls, this downpour must have caused damage; but, falling as it did into a basin of nearly 200 square miles in extent, with deteriorated channels and silted outfalls, the result can only be described as disastrous. From Hotar to Nāzrā the railway passed through an inland sea; and the tracts at a distance from the railway were equally congested. After the top of the flood had run off, these lakes still remained, and the water in the low-lying tracts, finding no exit, stayed where it was for months, until it evaporated. The rice-crops were destroyed and rotted in the water. The loss of crops was roughly estimated by the Executive Engineer at nearly 86 lakhs of rupees; and the loss of houses and cattle must have been very large.

A report on the flood was submitted by Mr. Maconchy, then Executive Engineer, Northern Drainage and Embankment Division, which showed that for the whole area of 717 square miles the existing ventage aggregated only 1,873 square feet, or less than half of what was required. Subsequently, in March 1902, the Executive Engineer made a thorough exploration of the great Jaynagar swamps and the country lying to the east of them. The need of drainage in this tract had been brought to notice in September 1899 by a petition for the construction of a sluice at Dhosā (near Habkā), but it had been treated as a drainage basin by itself, distinct from the great central tracts. The result of Mr. Maconchy's inquiries led him to make a strong recommendation that both the Habkā scheme and a scheme for reconstructing the Surjipur sluice at the mouth of the *khal*, on the bank of the Piāli river, should be included in the large Magrā Hāt scheme, on the ground that the drainage systems of all these tracts were so intimately connected with that of the central tract that they could not properly be treated separately. The Collector laid both the Magrā Hāt and Habkā schemes before a meeting of the District Board held on the 21st July 1903, and it was unanimously resolved that the Habkā scheme be incorporated in the Magrā Hāt scheme, and that the Government be moved to issue orders for the appointment of Drainage Commissioners under section 3 of the Bengal Sanitary Drainage Act. These were appointed, detailed plans and estimates were prepared, and in 1905 the scheme was adopted by the District Board and sanctioned by Government. Work was actually commenced in November 1904.

The scheme has been undertaken under the Bengal Sanitary Act, by the provisions of which the initial expenditure, together with the capitalized cost of maintenance, may be recovered from the proprietors of the tract affected. The estimated cost was a little over 20 lakhs, towards which Government made a contribution of 5 lakhs; it also undertook to bear all maintenance charges in return for any income derivable from the scheme, which includes such tolls as may be levied under the Canals Act for the use of navigable channels. The remainder of the cost has been met from a loan granted by Government to the District Board bearing interest at the rate of 4 per cent. per annum and repayable in 30 years, this being the maximum period fixed by the Act. In order to relieve the District Board as much as possible, the loan has been treated as an advance, and the District Board has been allowed to defer payment of interest and repayment of capital until recovery is being made from the proprietors. The work is to be maintained by Government, which will receive any navigation tolls and fishery rents that may be realized.

The scheme consists of three sections, viz., the Magrā Hāt section with an area of 219 square miles, the Habkā or Dhosā section (50 square miles) and the Surjipur section (20 square miles). The following is a brief general description of the works:—

The main sluice at Diamond Harbour is designed to discharge the drainage from an area of 215 square miles, including the whole of the areas drained by the Kāorāpukur Khāl; the tracts lying around Hotar, Nainān, Magrā Hāt, and Surjipur, which used to find an outlet through the Surjipur Khāl; the whole of the tract connected with the silted-up Sangrāmpur Khāl, as well as the area lying to the north of the railway between Magrā Hāt and the Nāzrā Khāl; the country to the north, which is drained by the Srīchandra Khāl; and the great Jaynagar swamp. Besides these, the main sluice has to accommodate the drainage discharging into the creek below Usti. The new Surjipur sluice is merely an extra outlet to relieve the main sluice; while the whole of the Habkā basin of 50 square miles discharges eastwards through the Habkā sluice.

Two main outfall channels are provided to feed the main sluice, both discharging into the creek; one from Nainān to Usti and the other following approximately the line of the old Sangrāmpur Khāl. Provision is made for discharging the Jaynagar swamps through the latter channel, while the water from the Surjipur swamp will be led to Usti. The width of the channels varies from 220 feet for the main creek leading to Diamond Harbour to 10 feet for the smallest channels, the sizes of the larger channels being determined by the probable flood discharge of storm water, and of the smaller by the requirements of navigation.

OTHER SCHEMES.

In addition to the above, a sluice has been constructed in the 108th mile of the Sundarbans embankment at Dhutkhāli, which drains 8 square miles (the Dhutkhāli basin), and a small sluice at Phulbāgichā in the 131st mile drains one square mile of the Khāri basin. Drainage schemes have also been proposed for the Kātākhāli, Khāri, Mahāmāyā, Arāpanch and Faltā basins. These and other projects are referred to as follows in a note prepared by the Executive Engineer in 1906:—

"A scheme for draining the Arāpanch basin of 30 square miles is under preparation, and work has been started in excavating the outer channel of the existing sluice. To the west of the Calcutta-Diamond Harbour road there are seven basins (shown in the margin), totalling 112 square miles, which require more sluice ventage. These basins are in urgent need of proper drainage facilities. At

Basin.	Area in sq. miles.
Munikhāli ...	16
Rāypur ...	22
Godākhāli ...	19
Kātākhāli ...	28
Faltā	18
Nainān ...	3
Nilā	6

present, drainage is effected through open *khāls* into the Hooghly, and these are very badly silted up. There are large inland swamps in which stagnant water remains till the beginning of the hot weather, when it evaporates. Instead of draining each basin separately, it might be possible to combine several basins together into one having one or perhaps two outfall sluices and a proper system of inland channels. The basins round Faltā might be combined in this way and a scheme carried out under the Sanitary Drainage Act. . . To the east and south of the Diamond Harbour road, and to the west of the Piāli river, matters will be much better when the present schemes being carried out are completed. The only basins not properly drained will be those shown in the margin: of these the Hārā, Jhinkrā, Khāri, Mahāmāyā, Dihi and Kāorāpukur basins are in urgent need of extra sluices.

Basin.	Area in sq. miles.
Hārā ...	... 24
Jhinkrā	... 8
Khāri ...	... 31
Mahāmāyā	... 5
Dihi ...	... 2
Kāorāpukur	... 28

"Hārā is an inland basin with no means of drainage. A sluice close to the existing one-vented sluice at Hārā is badly required, but the zamīndārs do not appear inclined to come forward and pay the amount necessary for the scheme. Practically no crops have been grown in this basin for years.

"The Khāri and Mahāmāyā basins might be combined. The total area is 36 square miles, and a sluice ventage of 180 square feet is required against 81 square feet existing. This basin is badly flooded by overflow from other basins, especially from the Jaynagar swamps now being drained by the Magrā Hāt scheme; the sluice is situated about 4 miles up the outfall *khāl*, which is badly silted. A main outfall sluice is required at the mouth and the *khāl* silt cleared.

"The original area of the Kāorāpukur basin was 52 square miles; 24 square miles have been included in the Māgrā Hāt scheme, leaving 28 square miles undrained. There is difficulty in draining this basin; the natural outfall is into Tolly's Nallah near Russā *viâ* the Kāorāpukur Khāl. This *khāl* is badly silted by the tides entering Tolly's Nullah . . . The drainage of the Jhinkrā basin does not seem urgent. The Dhutkhāli sluice has 80 square feet ventage for 8 square miles, or about twice as much as is required; this sluice will relieve the Sātpukur and Tāngrābichi sluices which should then be able to discharge the drainage of the Jhinkrā basin . . . In the south of the area near Satpukur the country is at present provided with sufficient sluice ventage, but the efficiency of the sluices, especially the Sātpukur sluice, is being gradually diminished owing to the silting up of the outfalls and creeks. The Subdivisional Officer, who is experienced in drainage schemes in this part of the division, reports that the deterioration of these channels is due to the reclamation of the Sundarbans. Areas are being reclaimed and the tidal spill reduced, diminishing the scour in the creeks and causing rapid silting.

"There is one other case which requires early attention Between Gariā and Surjipur there is an old river bed known as the Marāgangā Channel or Gangā Nadī; the channel is of practically no use for drainage purposes, and consists of large pools of stagnant water which do not entirely dry up till the very end of the hot weather. Many of the pools have been formed by excavation for raising the land on either bank, and others have been formed by cross-bunding the nullah. The channel passes through the Rājāpur and Bāruipur Municipalities and is used for depositing corpses and rubbish. Starting from near Gariā it passes south-eastward and runs parallel with the Gariā-Bāruipur road, crossing the Eastern Bengal State Railway about a mile to the south of Bāruipur railway station. Continuing in a south-easterly direction, more or less parallel to the Bāruipur-Surjipur road, it ends at Surjipur, the total length being about 15 or 16 miles. The channel at the northern end for about 5 miles is outside the Magrā area, and for the rest of the way forms the north-eastern boundary of the area to be drained by the scheme. The latter portion passes through comparatively high ground, and the general slope of the country is to the south. For this reason, the area to the south of the channel will be drained by the Hotar Khāl, which passes through the heart of the main swamp in the portion of the area to be drained. The Hotar Khāl falls into the Mārāgangā Channel, about 2 miles to the north-west of Surjipur, and this portion will be excavated under the scheme to form an outfall for the Hotar Khāl."

EMBANKMENTS. The Public Works Department maintains over 200 miles of embankments, of which all but a few miles were constructed, and are kept in repair, at Government expense. The main embankment runs southwards along the left bank of the Hooghly river from Akrā a few miles below Calcutta to Rāngāfalā near the head of Sāgar Island; thence it branches east and north to Sāmukpotā, and terminates at Gariā, 8 miles south of Calcutta. This embankment has a total length of 212 miles and protects a tract of 717 square miles in the south-west of the district from inundation by the Hooghly and Sundarbans rivers. Drainage is provided for by numerous sluices, of which thirteen are on a large scale.

The following is a statement of the principal sections of embankments maintained by Government:—

Embankment.	Length. Miles.	Length. Feet.
Hooghly left embankment	77	1,840
Sundarbans *id.* *id.*	53	2,640
Srīrāmpur Khāl right embankment ...	6	2,640
id. *id.* left *id.*	9	2,640
Khāri Khāl right embankment	3	602
id. *id.* left *id.*	3	2,040
Surjipur right embankment	8	...
id. left *id.*	4	2,640
Piāli river right *id.*	12	2,160
id. *id.* left *id.*	11	1,280
Embankment at the mouth of the Bāghmāri Khāl.	...	2,900
Bidyādharī river right embankment ...	10	3,120
Tolly's Nullah south *id.* ...	10	...
id. *id.* north *id.* ...	2	4,020
id. *id.* south *id.* (*taccāvi*) ...	2	1,320
Panchānnagrām embankment	3	1,400

CHAPTER VII.

NATURAL CALAMITIES.

LIABILITY TO CALAMITIES.

THE chief crop of the district is *āman*, or winter rice, which is grown on so large a proportion of the cultivated area, that the outturn of the *aus*, or early rice, however good, cannot compensate for its loss. For its successful cultivation the rainfall must be not only heavy, but also seasonable and well distributed; but fortunately it is rarely so deficient or badly distributed as to cause any serious or widespread failure. On the contrary, the chief danger to which the cultivators are exposed is excessive precipitation, resulting in prolonged inundations and the consequent destruction of the paddy seedlings. The natural and artificial means of communication, such as roads, railways, rivers and navigable creeks, are, however, ample to ensure the easy importation of grain to areas where there is a local shortage, and the people generally may be regarded as immune from famine. There is, however, no safeguarding against the sudden fury of a cyclone and the even more destructive storm-wave which sometimes accompanies it. To such cyclones the district is peculiarly exposed on account of its position at the head of the Bay of Bengal, and its records show that, though they occur at irregular intervals, these violent storms are far more destructive of life and property than either droughts or floods.

FAMINES.

Since 1770 the only famine from which the district has suffered is that of 1866, which, however, did not affect it very seriously. While it lasted, the maximum price of the rice in ordinary use was Rs. 6 a maund, and of paddy Rs. 2-8 a maund: even in localities where the distress was sharpest the price did not rise higher than seven seers a rupee. The first symptoms of distress were noticed in October 1865, at the end of which it was reported that "there were very palpable signs of great distress amongst the people, in consequence of the failure of the rice crop owing to the want of rain: it is impossible to disguise the fact that the distress which now prevails is only the commencement of what promises to be a famine." In some places the rice crop had entirely failed; rice

Famine of 1866.

was selling all over the district at double the rates which prevailed at the same time in the preceding year, and the price was rising rapidly. It was estimated that, if rain were to fall, the outturn of the crop would not be above half that of an ordinary year; if no rain fell, the outturn would not be above one-eighth. The result of the inquiries made by the Magistrate was that a failure of half the crop was expected; and it was found that in some places the people were selling their ornaments and brass vessels.

In May 1866, the Collector was authorized to begin relief works on certain roads in the south of the district if any need for employment should arise. Acute and general distress first made itself apparent in that month; a great increase in crime in the south of the district rendered additional police necessary. An inquiry into the state of the Diamond Harbour subdivision disclosed very real suffering; many people were living on leaves and roots, and the grain which had been reserved for sowing was being used for food. Money and rice were freely distributed to the famishing and relief committees were organized in July. Next month it became necessary to import rice into the district, 600 maunds of rice being bought at Kustia in Nadia and sent to Kulpi, south of Diamond Harbour, through the Sundarbans.

The Commissioner of the Division proceeded on a tour of inspection through the affected area in October, and reported as follows:—"From what we heard and saw it might be generally stated that throughout this tract the classes who ordinarily live by daily labour and wages, as well as those who live on alms, are now subsisting, not on rice, but on the roots of *bankachu* (a kind of wild yam) and the leaves of the *sajina*, tamarind and other trees boiled down. Nevertheless there was not that universal appearance of attenuation among the people which might have been expected. Almost every labouring man whom we met complained that no work was to be got; the prospect of work was universally received with delight, and with an urgent request that it might be immediate." The Commissioner directed the immediate commencement of works at as many places as possible, on three or four lines of road running through the distressed tract, and authorized the Magistrate to undertake any other work which he could devise. In the meantime the Public Works Department had supplied employment, for all who wanted it, on the embankments and roads of the Diamond Harbour subdivision. In the week ending 14th July, 2,360 persons were daily at work. The rains, however, brought the embankment repairs to a close, and the number of labourers employed by the Public Works Department fell gradually to about 300 daily, at which number it remained throughout the rainy season. The discontinuance of the works enhanced the distress at first; but the reaping of the early rice crop gave employment for some weeks, and such work as the Relief Committee could provide was going on. The cessation of agricultural operations in September, however, threw the people out of employment again, and an increase in the number of applicants for gratuitous relief immediately took place. But a general fall in the price of grain occurred in November, in consequence of importations from the eastern districts, and of some of the local cold weather crop being already in the market. During this month operations were contracted at the relief centres, and on the 24th November six centres in the tract immediately south of Calcutta were closed altogether.

The number of relief centres throughout the district was nineteen, and the numbers relieved on the 1st day of each month were as follows:—1st July, 264; 1st August, 1,162; 1st September, 3,156; 1st October, 8,862; and 1st November, 9,490. After this the number fell with extraordinary rapidity, till all relief was discontinued on the 1st December. Employment was given on nine roads over an aggregate length of 34 miles, the aggregate daily total of those thus employed being 31,876. The mortality from direct starvation was very small: only in two localities were deaths of residents of the district reported to have been directly attributable to starvation. In this, as in other districts, however, the high price of food reduced people to a condition in which they readily succumbed to attacks of diarrhœa and dysentery. The instances in which death was directly due to emaciation and want of food were mostly among people from Orissa and Midnapore, who came across the Hooghly on their way to Calcutta, or who wandered about the district in search of employment. Many of these were in such a state of emaciation when they arrived, as to be beyond the hope of recovery.

FLOODS. Floods occurred in 1823, 1838, 1856, 1864, 1868 and 1871 on such a scale as to seriously affect the crops of the district, but not such as to cause a general destruction of them. The flood of 1871, which was the result of excessive rainfall, was a serious inundation in the eastern and north-eastern portions of the district. In the inundated tracts, a large portion of the *āman* rice crop was destroyed; and when the waters subsided, a large number of cattle died, partly from want of food, and partly from eating grass which had become rotten from long immersion in water. The southern portions of the district, however, benefited by the large amount of fresh water brought down, and produced a bumper crop.

In recent years the most serious floods have been those of 1900, 1904 and 1907. In September 1900 there was abnormal rainfall, the total for the month being 38 inches, which caused floods all over the district and damaged nearly the whole of the standing crops. The damage was greatest in the ill-drained area between Diamond Harbour subdivision, where the accumulated water did not subside for months, and affected even the outturn of 1901. There was again an absolute failure of the crops in a limited area owing to excessive rainfall in June and July 1904, in consequence of which the low-lying lands were inundated and winter rice could not be grown in them. In 1905 the rainfall was not only heavy, but also uneven in its distribution. The low-lying lands in the south of the Basirhāt and Diamond Harbour subdivisions were inundated at the time of transplantation, and the standing crops were damaged. Next year the rainfall was neither sufficient nor evenly distributed, so that poor harvests were reaped; and in 1907 excessive rain in June and July damaged the jute and rice crops in the low lands, while on high lands the latter crop suffered later in the year on account of scanty rain in October and November. The result was distress, to cope with which the distribution of agricultural loans and other relief measures were necessary.

The monsoon in this part of Bengal consists of a series of cyclonic depressions, which follow each other in more or less close succession up the Bay of Bengal. Each period of general and heavy rainfall is, in fact, initiated by the advance of a cyclonic storm, which gives concentrated rainfall over long narrowish belts of country. In popular parlance, however, the term "cyclone" is reserved for the most violent storms which burst more especially in the transition periods, *i.e.*, in May before the monsoon is fully established, and in October when it has not altogether disappeared. "The conditions are, on the whole, favourable for the formation of the most severe cyclones in October, when the humid currents that provide the motive power are still of moderate strength. These October cyclones are examples of the most intense tropical storms. They differ in several respects from the cyclonic storms of temperate regions. They are usually of small extent, occasionally not exceeding 200 miles in diameter. Hence the shifts of wind accompanying them are very rapid and dangerous to vessels, and October is the most critical period of navigation in the Bay of Bengal. The pressure gradients are very steep, and the winds of hurricane intensity in the inner storm area. The precipitation is excessive, the rain being commonly described as 'falling in torrents.' The most characteristic feature in the worst storms is an inner central area of calms or light variable winds, occasionally 10 to 20 or even 30 miles in diameter, which is termed by sailors 'the eye of the storm.' The transition from the calm area to the belt of hurricane winds is usually exceedingly rapid.

"Another characteristic feature of these cyclones is the piling up of a mass of water in the inner storm area and area of lowest pressure. This advances with the storm and strikes the coast as a 'storm-wave.' The effect of this in flooding the coast districts depends largely upon the phase of the ordinary tidal wave at the time when the storm-wave strikes the coast. If the storm-wave strikes the coast about high water or shortly after, it may produce the most disastrous results, flooding low coast districts in a few minutes to a depth of 10, 20, or even 30 feet above tidal high water level. In such an inundation, caused by a storm-wave which spread up the Hooghly in 1737, 300,000 people are said to have perished, but the number is probably exaggerated. The storm-wave accompanying the Calcutta cyclone of October 1864 drowned 50,000 people and caused immense destruction of shipping. Cyclones of the most dangerous type are fortunately rare, not more than one, on the average, occurring in five years."*

Cyclone of 1737.

The earliest cyclone of which there is an historic account appears to have been that of 1582 A. D., which swept over Sarkar Bakla, *i.e.*, Backergunge, causing the loss of 200,000 lives; in this case also the destruction appears to have been caused mainly by a storm-wave.† It is not known whether this cyclone extended as far west as the 24-Parganas, but there is no such doubt about the cyclone of 1737 alluded to above. "Good God," wrote Sir Francis Russell, "what a sight was the town and the river in the morning! Not a ship but the *Duke of Dorsett* to be seen in the river, where the evening before were twenty-nine sail of vessels great and small, many being driven ashore, some broke to pieces, and others foundered. And this, which is scarce creditable in a river hardly a mile wide, there was no ebb-tide for near twenty-four hours. Our church steeple was blown down, as also eight or ten English houses, and numbers belonging to the black merchants. The whole place looked like a place that had been bombarded by an enemy. Such a

---

* *Imperial Gazetteer of India* (1907), Vol. I, pp. 134-5.

† *Ain-i-Akbari* and *Kiyazu-s-Salatin*.

havoc did it make that it is impossible to find words to express it."* A graphic account of this calamity, which also has some elements of humour, is given in the *Gentleman's Magazine* of 1738:—"On the 30th September last happened a furious hurricane in the Bay of Bengal, attended with a very heavy rain which rained 15 inches of water in 5 hours, and a violent earthquake which threw down abundance of houses; as the storm reached 60 leagues up the river, it is computed that 20,000 ships, barks, sloops, boats, canoes, etc., have been cast away. A prodigious quantity of cattle of all sorts, a great many tigers, and several rhinoceroses were drowned; even a great many caymans were stifled by the furious agitation of the waters. Two English ships of 500 tons were thrown into a village about 200 fathoms from the bed of the river Ganges, broke to pieces, and the people drowned pell-mell amongst the inhabitants and cattle. Barks of 60 tons were blown two leagues up the land over the tops of high trees. The water rose, in all, 40 feet higher than usual. . . A French ship was drove on shore and bulged. After the wind and water abated, they opened the hatches and took out several bales of merchandize, etc., but the man who was in the hold to sling the bales suddenly ceased working, nor by calling him could they get any reply. On which, they sent down another, but heard nothing of him, which very much added to their fear, so that for some time no one would venture down. At length, one more hardy than the rest went down and became silent and inactive as the two former to the astonishment of all. They then agreed by lights to look down into the hold, which had a great quantity of water in it, and to their great surprise they saw a great alligator staring as expecting more prey. It had come in through a hole in the ship's side, and it was with difficulty that they killed it, when they found the three men in the creature's belly."

Cyclone of 1864.

The most disastrous cyclone within living memory is that of 1864. The storm, which had been slowly travelling up the Bay of Bengal, made itself felt at the Sandheads on the afternoon of the 4th October and attained its full fury in the night. At Calcutta it raged from 10 A.M. till 4 P.M. on the 5th, after which it gradually subsided; here the lowest reading of the barometer was 28·571 at 2-45 P.M. The destruction caused by the cyclone was twofold. First, the violence of the wind caused widespread destruction to houses and trees. Secondly, the storm-wave brought up by the gale swept over the country to a distance

* *A History of Old Fort William in Bengal,* Bengal, Past and Present Vol. I, p. 44. The church referred to was St. Anne's.

of 8 miles inland on either side of the Hooghly as far north as Achipur. This wave rose in some places to a height of 30 feet, sweeping over the strongest embankments, flooding the crops with salt-water and carrying away entire villages. At Sāgar Island it was 15 feet above land level, and appeared to cut a channel straight across the island, dividing it into two halves. The embankments, houses, huts, *golas* and buildings were destroyed; and, out of a population of nearly 6,000, less than 1,500 survived. Those that did escape were saved by climbing up trees, or floating on the roofs of their houses, which the wave swept away and carried many miles inland. At Diamond Harbour the wave was 11 feet high, and it was stated at the time that it was impossible to go 50 yards on the road, at any place within six miles of Diamond Harbour, without seeing a corpse. Other villages on either side of the river suffered more or less: in some every house was swept away with most of the inhabitants. The distress and suffering to which the survivors in the affected tracts were exposed after the disaster were very great. For several days food was not obtainable, for the local stores had been swept away, and relief could not be sent from Calcutta. In some places which escaped the storm-wave the stores of the rice merchants were broken open and plundered; in others a kind of grass was eaten as food.

The cyclone wrought havoc among the shipping in the river. On the 5th October there were 195 vessels within the limits of the Calcutta Port. They withstood the force of the wind with success; but when to this, at about 1 P.M., was added the storm-wave, the force of which was still not entirely spent, one vessel after another broke from her moorings, and as each ship was swept on, she fouled others in her course. Massed together in hopeless and inextricable confusion, they were driven in heaps on the Sumatra Sand and along the Howrah shore from Sibpur to Ghoosery: there was, it must be remembered, no bridge between Calcutta and Howrah in 1864. Ten vessels were sunk in the river and 145 driven on shore. The *Govindpore*, a new ship of 1,200 tons, capsized and sank off the Custom House: the crew were saved by the gallantry of a sailor who swam off to the wreck with a line, by means of which the crew clinging to her masts escaped to shore. The *Ally* met the gale a little below Diamond Harbour; she had on board 348 cooly emigrants for Mauritius, and went down with all on board save seven of the crew and 22 emigrants. Six tug steamers were lost. The P. and O. Co.'s *Hindostan*, an old hulk, broke loose, turned over and went down off Garden Reach: their mail steamer *Bengal* stranded on the opposite side of the river, but was got off without serious damage: the Burma mail-steamer foundered off the Sandheads with nearly all hands: a hospital ship was carried on to the top of the Diamond Harbour embankment: two light-ships were lost with all hands. On land very extensive injury was inflicted on the public works and buildings. At 6 P.M. the Strand Road was flooded throughout, and in places the water stood breast high. The avenues in Fort William and the Botanic Garden were destroyed: the Eden Gardens were turned into a wilderness: the Barrackpore Park lost 50 per cent. of its valuable trees, and the avenue on the Barrackpore road suffered even more.*

Part of the district was again visited by a cyclone on the 1st November 1867, the centre of the storm traversing the country nearly due east from Calcutta to Basirhāt on the Ichāmatī river. In this line villages were blown down wholesale, and their destruction was accompanied by loss of human life, the more populous places which suffered severely being Bāruipur, Diamond Harbour, Basirhāt and Gobardānga. The effects of the hurricane were most disastrous in Port Canning, where the gale was accompanied by a storm-wave, the water of which passed over the town with fearful violence. The station-house, goods' sheds and railway hotel were all blown down and the Port Canning Company's store hulk carried away a large portion of the railway jetty. The storm-wave, beginning from Sāgar Island, extended to the extreme east of the district, and in some rivers the water rose to 6 feet above flood level. Cyclone of 1867.

The limits of space preclude a detailed account of other cyclones. Reference must however be made to a cyclone which swept over Sāgar Island in May 1833. In June 1823, only ten years previously, a storm had destroyed the roads, embankments and crops on this island, but the cyclone of 1833 was even more terrible. The island was submerged to a depth of 10 feet, and the whole population of 3,000 to 4,000 souls is said to have perished. On this occasion an East Indiaman, the *Duke of York*, was carried into the rice fields at Falta and left there high and dry. A curious incident which occurred during a storm that burst in the Hooghly in May 1893 may here be mentioned. A large German steamer went aground on a sand bank, and a number of lighters were sent in the hope that she might be got off by taking out part of her cargo. While they were busy lightening her, the tide rose, and a second small cyclone came on, which Other cyclones.

* C. E. Buckland, *Bengal under the Lieutenant-Governors*, Vol. I, pp. 298-300.

blew the steamer back into the river uninjured, but sank several of the lighters.

The last cyclone which has visited the district occurred on 17th October 1909, but was severe only in the Basirhāt and Bārāsat subdivisions, where it caused considerable damage to houses, trees, crops and cattle. The storm raged from 2 P.M. on the 17th to 2 A.M. on the 18th, and it is reported that the wind blew from the north-east at Basirhāt, from the north at Bārāsat, from the north-west at Barrackpore, and from the south-west at Diamond Harbour; in the intervening space its direction varied from between north-east and north-west. It may not be out of place to mention that the motion of the wind in a cyclone is in an involute spiral, revolving in a direction opposite to that of the hands of a clock.

EARTH-QUAKES.

The severest earthquake within the memory of the present generation occurred on 12th June 1897, when many buildings were damaged and others brought down. In Calcutta the steeple of the Cathedral was destroyed and 1,300 houses were injured. Another earthquake was experienced on 14th July 1885, and there were several earth-tremors of less severity in previous years of the eighteenth century.

CHAPTER VIII.

RENTS, WAGES AND PRICES.

THE rents paid by the actual cultivators to their immediate landlords vary according to the nature and quality of the land, but the general incidence is high. Rates are highest in the Sadar and Bārāset subdivisions, where rice lands are rented at Rs. 6 to Rs. 12 an acre; even higher rates prevail for homestead and sugarcane lands, which fetch Rs. 10 to Rs. 20 an acre. The following are reported to be the average rents for different classes of land in each subdivision:— RENTS.

SUBDIVISION.	RATES PER ACRE.				
	Homestead land.	*Aman* land.	*Aus* or *jot* land.	Orchard land.	Sugarcane land.
	Rs.	Rs.	Rs.	Rs.	Rs.
Sadar	15	9	15	15	...
Bārāset	10—20	6—12	6—15	9—15	...
Basīrhāt	6—12	3—15	18—24	6—12	18—24
Diamond Harbour ...	3—10½	3—10½	3—10½	3—10½	3—10½

In the Falta and Diamond Harbour thanas of the Diamond Harbour subdivision the rents paid by *korfa* or *thika* raiyats, who are tenants-at-will, are much higher than those shown in the table, ranging from Rs. 15 to Rs. 36 per acre. In the Sundarbans the usual rate of rent is Rs. 6 to Rs. 7-8, while in the Basīrhāt subdivision the rate for jute land rises to Rs. 12, and that for *pān* and betel gardens to Rs. 18 per acre.

Produce rents are paid as well as cash rents, the system being known as *bhāg chās* or *bhāg jot*; under this system the cultivators usually pay their landlords half the produce of their fields.

The annual profit of cultivation is estimated at Rs. 17-4 per *bigha* of rice land, exclusive of rent, the basis of the estimate Profits of cultivation.

being as follows: The cost of preparing a *bigha* of land as a nursery for seedlings is taken at Rs. 5, viz.—(1) Rs. 3 for three ploughings with two ploughs at a time, *i.e.*, six ploughs at 8 annas a plough; (2) Re. 1-4, the cost of 15 *pālis* (30 seers) of seeds; (3) 12 annas for weeding, two days' labour of one man. The seedlings on one *bigha* will suffice for the transplantation of 10 *bighas* of rice land, so that the cost of seedlings per *bigha* comes to 8 annas only. The cost of cultivation of paddy land may be taken at Rs. 8-12, viz.—(1) 8 annas, being the cost of the seedlings, as above stated; (2) Rs. 3 for three ploughings, with two ploughs at a time, at 8 annas per plough; (3) Re 1-8 for transplantation, including the removal of seedlings from the nursery, 4 labourers at 6 annas a day; (4) 12 annas for weeding, 2 labourers at 6 annas a day; (5) Re. 1-8 for reaping, 4 labourers at 6 annas a day; (6) Re. 1-8 for conveying the crop to the threshing floor and for threshing, 4 labourers at 6 annas each. The outturn per *bigha* is taken at 5 maunds of clean rice and one *kāhan* of straw, which will fetch Rs. 26, viz., Rs. 20 for the rice and Rs. 6 for the straw. If, therefore, Rs. 8-12 be deducted for the cost of cultivation, there is a balance of Rs. 17-4; and taking Rs. 3 as the average rent of a *bigha* of rice land the net profit comes to Rs. 14-4. An ordinary cultivator holding 15 *bighas*, or 5 acres, will consequently have an annual income of Rs 213-12, assuming that he employs hired labour. As a matter of fact, however, a man with a holding of this size usually cultivates his lands himself, except at the reaping and transplanting seasons, when he has to employ labourers. About half of the cost of cultivation will be saved if he and his family contribute their share of manual labour, so that his net profit per *bigha* will be Rs. 18-10 and his annual income will amount to Rs. 279-6.

Size of holdings.

The district having had no settlement, accompanied by the preparation of a record-of-rights, it is impossible to state with any degree of accuracy the average size of the cultivators' holdings. The majority appear to have tenancies not exceeding 5 to 6 acres, and the general average is probably from 3 to 4 acres. In the Government estates in the Diamond Harbour subdivision the average varies from 2½ to 17, acres as shown in the margin; Belpukur, Kedua and Paila Bhagwānpur are situated near the Sundarbans, where tenancies are larger than elsewhere.

Estate.	Acres.
Kāmārpol	2½
Banamālipur	4½
Belpukur	7⅓
Kedua	8
Paila Bhagwānpur ...	17½

WAGES.

The marginal table showing the daily wages paid for different classes of labour during the last fortnight of March sufficiently illustrates the tendency of wages during the last 20 years. Though the custom of paying village artizans and menials in kind appears to have fallen into desuetude in the Sadar and Diamond Harbour subdivisions, it still lingers in parts of the Bārāset and Basirhât subdivisions. Here a village blacksmith gets one maund of paddy annually from each of the villagers whose ploughshares he makes and repairs, and the village barbers and washermen are similarly remunerated by a share of the harvest. In the *ābāds*, or reclaimed portions of the Sundarbans, the reapers either receive wages in cash at the rate of Rs. 7-8 a month, with two meals a day, or in kind, being given one or more bundles of paddy for every 20 bundles that they cut; eight or ten of these bundles will yield 2½ seers of paddy. These reapers, who are called *daulias*, annually migrate to the Sundarbans for the rice harvest and return to their homes when it has been reaped.

Class of labour.	1893.	1903.	1912.
	Rs. A.	As.	Rs. A.
Superior mason ...	1 0	12	1 0
Common „ ...	0 8	9	0 10
Superior carpenter ...	1 0	12	1 0
Common „ ...	0 8	9	0 12
Superior blacksmith	0 10⅔	12	1 0
Common „ ..	0 8	9	0 10
Cooly (male) ...	0 4½	4	0 5¼
„ (female) ...	0 4	2	0 3½

In the mills and factories there are special rates for different classes of skilled and unskilled labour. The following table shows the *lowest* monthly wages paid during 1911 in the more important manufactories.

CLASS OF LABOUR.	JUTE MILLS		Jute presses.	Cotton spinning mills.
	Budge-Budge.	Gauripur.		
	Rs.	Rs.	Rs.	Rs.
Engine-driver	19	32	12	23
Weaver (male)	19	20	...	15
Blacksmith	13	20	16	15½
Fitter	13	20	10	22
Boilerman	12	19	10	13
Carpentor	13	18	17	14½
Mason	13	...	17	...
Bricklayer	12	12	...	13
Spinner (male)	9	10	...	9
Dyer	8	10	...	...
Cooly (male)	8	10	...*	10
„ (female)	8	7	...	8

* Daily wages were paid, the lowest being 5 annas a day.

PRICES. Statistics of the prices paid for various articles of food in each subdivision during the last fortnight of March from 1893 to 1912 will be found in the B volume, which forms a statistical appendix to this volume. In recent years the level of prices has been highest in 1908, when common rice sold at 8 seers per rupee in the Sadar subdivision, at 6 seers 14 chittacks in Bārāsat, at 7 seers 3 chittacks in Basīrhāt, and at 5 seers 8 chittacks in Barrackpore. With these figures may be compared those for the famine year of 1866, when the maximum price of common rice, at the height of the distress, was 7 seers per rupee. Prices generally have risen during the present century, with one notable exception, the price of salt having fallen considerably owing to the reduction of the duty.

Weights and measures. The standard maund, of 40 seers or 82 lbs. avoirdupois, and the standard seer of 80 tolas are in universal use throughout the district in buying and selling by weight. Gold, silver, and precious stones are weighed thus: 4 *dhān* = 1 *rati*, or 1·875 grains Troy; 6 *rati* = 1 anna, 8 *rati* = 1 *māsa*, or 15 grains Troy; 12 *māsa* = 1 tola, or 180 grains Troy. For large articles the following weights are used:—5 tolas = 1 chittack, or 2 ozs.; 4 chittacks = 1 *poā*; 4 *poā* = 1 seer, or 2 lbs. 0 oz. 14 drs.; 5 seers = 1 *pasurī*; 8 *pasurī* or 40 seers = 1 maund of 82 lbs. avoirdupois.

The local measures vary in different parts of the district. In Bārāset subdivision the scale is as follows:—2½ seers = 1 *pālī*; 2 *pāli* = 1 *don*; 2 *don* = 1 *kāti*; 8 *kāti* = 1 *ārhi*; 20 *ārhi* = 1 *bis*; 16 *bis* = 1 *kāhan*. In Diamond Harbour:—2½ seers = 1 *pāli*; 4 *pāli* = 1 *kātha*; 5 *kātha* = 1 *pan*; 4 *pan* = 1 *sali*; 4 *sali* = 1 *kāhan*. In the south of Diamond Harbour the scale is:—2 seers and 6 tolas (or 166 tolas) = 1 *pāli*; 20 *pālis* = 1 *kurih*; 10 *kurihs* = 1 *bisi*; 16 *kurihs* = 1 *kāhan*. In parts of Basīrhāt 5¼ seers = 1 *pāli* and 16 *pālis* = 1 *bis*. In other places the following measures are found:—(1) 2½ seers = 1 *kātha* or *pāli*; 20 *pālis* = 1 *sali*; 16 *salis* = 1 *kāhan*; (2) 6 seers and 6 chittacks (or 1½ *poās*) = 1 *pāli*; 2 *pālis* = 1 *kāhan*; 4 *kāhans* = 1 *ārhi*; 20 *ārhis* = 1 *bis*. (3) 2½ *poās* (or 10 chittacks) = 1 *khanchi* or *kanki*; 2 *khanchis* = 1 *rek*; 2 *reks* = 1 *pāli*; 16 *pālis* = 1 maund. (4) 5 chittacks = 1 *kanika*; 4 *kanikas* = 1 *rek*; 4 *reks* = 1 *pāli*; 20 *pālis* = 1 *sāli*; 16 *sālis* = 1 *kāhan*.

The following measure of numbers is used for cowries, bundles of straw, mangoes, eggs, etc.:—4 units = 1 *gandā*; 5 *gandās* = 1 *buri*; 4 *buris* (or 20 *gandās*) = 1 *pan*; 16 *pans* = 1 *kāhan*. Milk is sometimes sold at 75 tolas per seer, but the standard seer of 80 tolas is generally used. A common weigh for the sale of small fish is the *kankia*, which is equivalent to 1½ chittacks. In selling cloth a yard is equal to 16 *girās* and a *girā* to 2¼ inches.

The unit of lineal and square land measure is the *hāth*, which may be anything from 18 to 29 inches, but is usually 18 inches. Using this as the unit, the usual scale is 4 *hāths* = 1 *kātha; 20 kāthas* = 1 *bigha. Amīns* use the following measures:—1 *karā* = 3 *krāntis* or 16 *bisas* or 80 *tils; 4 karās* = 1 *gandā; 5 gandās* = 1 *kāchchhā; 4 kāchchhā* = 1 chittack; 16 chittacks = 1 *kātha*; 20 *kāthas* = 1 *bigha*. There are in the district *bighas* varying from eighty to a hundred *hāths* square, of 18 inches each, and also one of fifty-five *hāths*, of 29 inches each. The value of each in English measurement is as follows:—1 *bigha* of 80 *hāths* of 18 inches = 0·330578 acre; a *bigha* of 85 *hāths* of 18 inches = 0·373192 acre; a *bigha* of 90 *hāths* of 18 inches = 0·418388 acre; a *bigha* of 93 *hāths* of 18 inches = 0·446746 acre; a *bigha* of 95 *hāths* of 18 inches = 0·466167 acre; a *bigha* of 100 *hāths* of 18 inches = 0·516529 acre; a *bigha* of 55 *hāths* of 29 inches = 0·405575 acre. The corresponding value of an acre in *bighas* is as follows:—1 acre = 3·025 *bighas* of 80 *hāths* of 18 inches; 1 acre = 2·67958 *bighas* of 85 *hāths* of 18 inches; 1 acre = 2·39012 *bighas* of 90 *hāths* of 18 inches; 1 acre = 2·23841 *bighas* of 93 *hāths* of 18 inches; 1 acre = 2·14515 *bighas*, of 95 *hāths* of 18 inches; 1 acre = 1·93600 *bighas* of 100 *hāths* of 18 inches; 1 acre = 2·46563 *bighas* of 55 *hāths* of 29 inches. The standard *bigha* is the one of 80 *hāths* of 18 inches square equal to 14,400 square feet, or within a fraction of a third of an English acre, and is the one now more generally used.

The *bigha* comprises (1) 80 *hāths* in parganas Māgura, Azimābād, Havilisahar, Khāspur and Garh; (2) 90 *hāths* in pargana Dakhin Sāgar; (3) 93 *hāths* in pargana Penchākuli; (4) 95 *hāths* in pargana Munragācha, and (5) 110 *hāths* in parganas Anwarpur, Balanda, Amīrābād, Bhaluka, Buran, Maihāti, Sarfrāzpur, Chaurāsi and part of Dhuliapur.

Time is subdivided as follows:—60 *anupal* = 1 *bipal; 60 bipal* = 1 *pal; 60 pal* = 1 *danda*, equal to 24 minutes; 7½ *danda* = 1 *prahar*, or three hours; 8 *prahar* = 1 *dibas*, or day and night of 24 hours; 15 *dibas* or days = 1 *paksha; 2 paksha* = 1 *mās* or month; 2 *mās* = 1 *ritu; 3 ritu* = 1 *ayan*, or half-year; 2 *ayan* = 1 *batsar*, or year.

CHAPTER IX.

INDUSTRIES, MANUFACTURES AND TRADE.

OCCUPATIONS.

THE statistics compiled from the returns of occupations which were made at the census of 1911 show that 1,603,000 persons, or two-thirds of the population, are supported by agriculture, 391,000 or 16 per cent. by industries, 212,000 or 9 per cent. by commerce (including transport whether by rail, road or river), and 44,000 or 2 per cent. by the professions and liberal arts.

Of those maintained by agriculture, 68,000 subsist by income derived from the rent of agricultural land, *i.e.*, consist of landlords and their families, 1,250,000 are cultivators, and 271,000 are farm servants and field labourers. Taking the figures for actual workers only, there are 17,000 landlords, 376,000 cultivators and 106,000 agricultural labourers: in other words, there are five landlords and 28 agricultural labourers to every 100 cultivators. As might be expected in a district with a large water area, in the form of rivers, creeks and swamps, pasturage provides a livelihood for comparatively few and fishing for comparatively many. Including all those supported by pasturage and cattle-keeping, whether as breeders or herdsmen, and also those who trade in milk, *ghī* and butter (for it is generally a matter of chance whether a man who keeps cows is returned as a cattle-keeper or a milkman), the total is only 19,000. The aggregate of those who subsist by fishing is more than thrice as great, viz., 63,000, of whom 40,000 were returned as fishermen and their families, and 23,000 as fish-dealers. The two groups may be taken as connoting the same occupation, for though some live by fishing only and others retail but do not catch fish, the great majority ell the fish which they catch.

Nearly half of these supported by industrial occupations subsist by textile industries (186,000). By far the most important textile industry is the pressing, spinning and weaving of jute, which accounts for 169,000 persons. Then come, *longo intervallo*, tailoring and dress making (34,000) and the domestic work of rice pounding and husking (30,000), which is carried on almost exclusively by women. Cotton spinning and weaving, once so important a factor in the economic life of the peasantry, now provide a means of livelihood for only 11,000 persons, of whom nearly 5,000 were at work in cotton mills at the time of the census, leaving the exiguous total of 6,000 for the cottage industry.

The aggregate of those classified under the head "Transport" is 72,000, of whom nearly half are dependent on the provision of transport by road, such as cart-owners, cart-drivers, *pālki*-bearers and their families. Those supported by work on the railways aggregate 18,000, and by traffic on the water 15,000, of whom 9,000 are boatmen and boat-owners.

Service in the public force, such as the army and police, and in different branches of the public administration, furnish altogether 22,000 persons with their daily bread. Twice as many are engaged, or are dependants of those engaged, in work connected with the professions and liberal arts. The returns for actual workers under this latter head are interesting as showing how small a fraction of the population are engaged in professional, artistic and scientific pursuits, either because they are not sufficiently well educated or because they are debarred by want of means, opportunity or training, or by traditional custom, from following them, or because they do not find them sufficiently attractive or lucrative. Nearly half of the workers in the professions and liberal arts consist of persons having some religious avocation, such as priests, religious mendicants, temple servants, etc., their actual number being 7,000. As regards other workers, those engaged in educational work number 2,970 and in medical pursuits 2,349, including midwives and nurses, as well as medical practitioners of all kinds, whatever their qualifications. The legal profession has only 834 adherents, including lawyers' clerks and touts in addition to barristers, pleaders and mukhtārs, while those who are grouped together under the head "Letters, Arts and Sciences" aggregate only 1,244. This latter figure cannot be regarded as a large one, considering that there are nearly 2½ million persons in the district, and that the head comprises a wide range of pursuits, *e.g.*, music, painting, acting, dancing, architecture, engineering, etc.

Domestic service provides for 34,000 persons, while the number of those living on private income is 7,000, and of those engaged in or dependent on unproductive pursuits, such as beggars and prostitutes, 17,595.

INDUSTRIAL CENSUS.

The statistics of occupation compiled from the returns made at the census while indicating the main functional distribution of the people furnish meagre information concerning individual industries and manufactures. To remedy this defect, an industrial census was held in 1911, concurrently with the general census, *i.e.*, the owners, managers and agents of industrial works, employing 20 persons or more, submitted returns in which, *inter alia*, the number of their employés at the date of the census was entered. These returns, of course, only refer to the state of affairs on that date, when some of the concerns were closed, others were not in full work, and others had a larger number of operatives than usual. But even so, they furnish valuable information regarding the industrial development of the district. Briefly, the result is to show that on the 10th March 1911 there were in the 24-Parganas 175 industrial works, each with 20 hands or more, in which altogether 161,638 persons were employed. Of these, 4,519 were engaged in direction, supervision and clerical work, 53,884 were skilled workmen and 103,235 were unskilled workmen. These figures include 860 Europeans and Anglo-Indians (the designation prescribed officially for those generally known as Eurasians), of whom 742 were managers, supervisors or clerks, and 118 were skilled workmen.

WORKS.	Number.	Number of employés.
Jute mills	31	116,216
Arms and ammunition factories.	5	8,713
Brick-fields and surki and tile manufactories.	36	7,915
Jute presses	20	5,371
Cotton mills	4	4,696
Dockyards	2	3,203
Railway workshops ...	1	3,183
Paper mills	2	2,894
Tin works	4	2,703
Oil mills	22	956
Lac factories	2	715
Sugar factories	1	617
Other works	45	5,456
Total ...	175	161,638

The principal industry of the district is the manufacture of jute, which, at the time of the industrial census, provided employment for altogether 121,587 persons, viz., 116,216 in jute mills and 5,371 in jute presses. The number of these and other works is shown in the marginal statement above, together with the number of employés.

FACTORY INDUSTRIES. The proximity of Calcutta, itself a large consumer, with its facilities for export over-seas, and the many means of communication with the interior, by both rail and river, have led to the establishment and development of important factory industries along the Hooghly from Budge-Budge to Naihāti. In 1911 there were no less than 122 factories coming within the operation of the Factories Act (*i.e.*, employing 50 hands or more) out of a total of 320 for the whole of Bengal, while the average daily number of operatives amounted to 164,000, or more than half the aggregate for the Province. The list of factory industries is a long one, for it includes jute spinning, weaving and pressing, cotton spinning and weaving, paper making, sugar refining, soap making, bone grinding, brick making, the manufacture of lac, ropes, etc. In addition to private undertakings, there are several important works under Government control, which manufacture arms and ammunition, clothing for the troops and telegraph stores. In recent years the *swadeshi* movement has led to the establishment of several new manufactories, more especially in the suburbs of Calcutta near the Circular and New Cut Canals, where tanneries, bone crushing mills, and factories for the manufacture of ink and Portland cement have sprung up. Saw mills and rope works have also been started by Indian enterprise, while the manufacture of umbrellas, tin boxes and steel trunks has been taken up.

The following is a brief account of the more important factories:—

Jute mills. The manufacture of jute into gunnies, as jute cloth is called, is an industry of comparatively recent creation but very rapid growth. There are according to the returns for 1911, altogether 57 jute mills in Bengal, which consume, on the average, fully half the total quantity of jute produced. Their consumption of the fibre has been practically doubled in the last ten years, and the manufacture of gunnies has been largely diverted from Dundee to the banks of the Hooghly : at present, the mills confine themselves to the production of the coarser classes of goods, chiefly gunny bags and hessian cloth. Of these mills, no less than 39, employing (in 1911) a daily average of 122,000 hands, are in the 24-Parganas, being situated along the Hooghly from Budge-Budge northwards to Gauripur. Jute manufacture appears to have been started in the Alipore Jail in 1868; the jail jute mill is still at work, its products being taken chiefly by other jails and the Supply and Transport Department of the Indian army. Private enterprise entered the field after 1873, the Olive Mill being opened in 1874, the Shāmnagar Mill in 1875, and the Budge-Budge and Kāmārhāti Mills in 1877.

Several more mills were opened in the next decade, such as the Union, Upper Hooghly, Kānkināra, Titāgarh and Soorah; but the majority are of more recent date. The following statement shows the mills at work in 1911 and the average daily number of operatives employed in each:—

NAME OF MILL.	Number of operatives.	NAME OF MILL.	Number of operatives.
Albion	3,251	Kāmārhāti, No. 1 ...	3,682
Alexander	1,870	Ditto, No. 2 ...	3,351
Alipore Jail	820	Kānkināra, No. 1 ...	3,683
Alliance, North ...	3,002	Ditto, No. 2 ...	1,175
Do., South ...	2,040	Kelvin	2,765
Anglo-India, Upper ...	2,937	Khardah, No. 1 ...	4,405
Ditto, Middle ...	4,629	Ditto, No. 2 ...	5,103
Ditto, Lower ...	2,685	Kinnison	5,339
Auckland	2,494	Lansdowne	4,046
Barānagar, North ...	3,550	Naihāti	2,336
Ditto, South ...	2,798	Reliance	3,117
Ditto, Branch ...	1,422	Shāmnagar No. 1 ...	4,547
Budge-Budge	6,942	Ditto, No. 2 ...	1,759
Clive, No. 1	2,917	Soorah	1,176
Do., No. 2	2,775	Standard	3,849
Fort Gloster	312	Titāgarh, No. 1 ...	5,637
Gauripur	8,445	Ditto, No. 2 ...	6,537
Hooghly, Lower ...	(Closed)	Union	2,803
Do., Upper ...	3,388		

Jute presses.

The pressing of jute by machinery into bales for export was started in 1873, when two press-houses were established, viz., the Calcutta Hydraulic Jute Press and the Cossipur Jute Warehouse: these are still at work. The latter, which is the property of Ralli Brothers, is the largest press-house in Bengal. In 1911 there were 19 presses at work, with a daily average of 8,939 operatives, as shown in the margin: another

NAME.	Number of operatives.
Ashcroft	320
Atlas	165
Bengal (Hydraulic)	471
Calcutta (Hydraulic)	320
Camperdown	450
Canal	325
Chitpur (Hydraulic)	410
Cossipur Jute Warehouse	2,300
Gunga (Jute Mill)	212
Hoare, Miller & Co.'s	180
Hooghly (Hydraulic)	270
Jheel	380
Lakhi	508
Ocean	218
Strand Bank	500
Sun	260
Suraj	900
Union	425
Victoria (Hydraulic)	325

press, called the Bhajorām Jute Press, was closed throughout the year. The jute presses are nearly all situated in the northern suburbs of Calcutta, *i.e.*, in the towns of Cossipur and Chitpur.

Jute baling is carried on at the Nārkaldānga Jute Works, which are also called the Bridge Jute Mill.

There were four cotton spinning mills at work in 1911, viz., the Bengal, which had a daily average of 2,000 operatives, the Dunbar with 910, the Dunbar Ring with 576, and the Empress of India with 677. The oldest of these mills is the Dunbar Cotton Mill, which was opened in 1875, and the Empress of India Mill, which started two years later. The articles produced are ruled yarn, folded yarn and woven goods, which are supplied to the Indian and China markets. Another mill, called the Deshi Weaving Mill, at which calico is woven, remained virtually closed throughout the year. **Cotton mills.**

There are two paper mills in the district, both of which belong to the Titāgarh Paper Mills Company, Limited. One is at Titāgarh, and is known as the Titāgarh Paper Mill No. 1; the other, which is called the Titāgarh Paper Mill No. 2, is at Kānkināra. The former employed 1,423 hands and the latter 1,256 in 1911; between them, they produce about 35 million lbs. of paper annually. They manufacture white printing, *bādāmi*, coloured printing, cartridge, blotting and foolscap paper. The raw materials used are rags, grass, straw, jute, old gunnies, hemp baggings, ropes and waste paper. **Paper mills.**

Government itself is a large employer of labour, having five important manufactories, four of which supply the army with arms, ammunition and clothing. These are:—(1) the Gun and Shell Factory at Cossipur, which manufactures ordnance fittings, shells, fuses, cartridge metal, etc.; it employed 1,271 hands in 1911. (2) The Ammunition Factory at Dum-Dum, which started work in 1846, and turns out the cartridges, small arms, etc., required by the Indian army. The Dum-Dum bullet, a soft-nosed bullet that expands and lacerates the object it strikes, is so called because it was manufactured here. The average number of men in the works was 2,681 in 1911. (3) The Rifle Factory at Ichāpur, which was erected on the site of an old gunpowder factory and started the manufacture of rifles for the Indian army in 1907; it employed 2,050 men in 1911. (4) The Army Clothing Factory at Alipore, which was established in its present position in 1852. This factory produces uniforms and other clothing for the army in India, and employed, on the average, 380 hands in 1911. The fifth industrial concern under Government management is the Telegraph Workshop at **Government works.**

Alipore, which was opened about 1859, and afforded employment to 639 men in 1911. Here the articles required for the telegraph system, from telegraph posts and cables to instruments, are manufactured and repaired.

Engineering works and foundries.

There are altogether 13 engineering works and foundries in the district as shown in the marginal statement, nearly all situated in the immediate vicinity of Calcutta. The largest of these is the Vulcan Iron Works which used to belong to Messrs. Parry & Co. and were removed to their present site (in Lower Circular Road) in 1899.

NAME.	Number of operatives, 1911.
Vulcan Iron Works	442
Victoria Engine Works	167
Sikdār and Co.'s Iron Foundry ...	88
Bengal Engineering Works or Hastings Bridge Works.	60
Phœnix Iron Works (Jessop's) ...	340
Incell and Silk's Works	150
Saxby and Farmer's Factory	260
Hooghly Docking and Engineering Co.'s Works	60
Albert Iron Works	60
East Bengal Engineering Works ...	55
Russa Engineering Works	140
Civil and Sanitary Engineering Co.'s Works	45
Calcutta Municipal Works	330

Railway and tramway workshops.

The workshops of the Loco. and Carriage Department of the Eastern Bengal State Railway are situated at Kānchrāpāra; these are large works, affording employment to 2,158 persons (in 1911). The workshops of the Calcutta Tramway Company, which employed 825 hands in the same year, are situated in Entally.

Docks and dockyards.

A large labour force is in constant employment at the docks in Kidderpore. These docks provide for the whole export trade of Calcutta; they were commenced in 1884 and declared open in 1892. The Royal Indian Marine Dockyard, which in 1911 employed 1,550 workmen, is also at Kidderpore. There are three other dockyards, viz., those of the Port Commissioners, with 757 hands, the India General Steam Navigation Company, with 1,331 hands, and the Rivers Steam Navigation Company, with 1,271; the last two are situated in Garden Reach.

Cigarette factories.

Three factories, owned and managed entirely by Indians have recently been started for the manufacture of cigarettes, and have attained a fair measure of prosperity. They use imported modern cigarette-making machines, and have a well organized system of distribution: the cheapest quality of cigarettes sell at ten for a pice. These factories belong to the Globe Cigarette

Co., the Calcutta Cigarette Co., and the East India Cigarette Manufacturing Company. The largest factory is that owned by the company last named, which had 487 workmen in 1911.

Pottery works.

Another undertaking, owned and managed by Indians, which is making good progress, is the Calcutta Pottery Works, which were started in 1909. The firm manufactures cheap cups and saucers, images of gods and goddesses, etc., which compare favourably with similar imported goods. The raw material is imported from Rājmahāl. These works, remarks Mr. J. G. Cumming in his *Review of the Industrial Position and Prospects in Bengal in 1908*, "are the result of the enterprise of Mahārājā Manindra Chandra Nandi of Cossimbazar, and Bābu Baikuntha Nāth Sen of Berhampore. This firm is a good example of what is required in Bengal for indigenous development. It has sufficient working capital; it has its own coal and its own raw material; it has a ceramic expert in Mr. S. Deb, who studied at the Higher Institute of Technology in Tokio, Japan; it has up-to-date German and English machinery, and up-to-date furnaces with the best of Stourbridge fire-clay bricks; it has found a local market, as well as an export market, in ink-pots, gallipots, insulators, cups, saucers and plates, and even dolls. It is a curious development to find in Calcutta, as the product of Indian labour, such a western product as a china-ware doll dressed in European clothes."

Chemical works.

Another enterprise conceived in a true *swadeshi* spirit is the Bengal and Pharmaceutical Works, which Mr. Cumming describes as "one of the most go-ahead young enterprises in Bengal. Dr. Prafulla Chandra Ray, D.SC., F.C.S., started it as a small private concern in Lower Circular Road about 15 years ago and made drugs from indigenous materials. About six years ago it was made into a limited liability company with a capital of two lakhs. Many of the leading chemists in Calcutta are share-holders. It has now a well-thought-out and well-managed factory with about 70 workmen* at 82, Mānicktollah Main Road. Bābu Rājshekhar Bose, the manager, is an M. A. in Chemistry. The variety of manufactures is considerable. Besides the manufacture of drugs and acids, the manufacture of laboratory apparatus, which requires skilled craftsmen in wood and metal, has been taken up. The latest development is in perfumes. The enterprise shows signs of resourcefulness and business

* The average daily number in 1911 was 190. The larger figure sufficiently demonstrates how the works have developed since 1908, when Mr. Cumming drew up his report.

capacity, which should be an object lesson to capitalists of this province."*

Tin works. At the bulk oil depôts established at Budge-Budge by the oil companies, such as the Burma, Standard and Asiatic Petroleum Companies, kerosine oil tins are manufactured. Twenty years ago not a single tin was made in Bengal, but now there are five factories, with modern stamping machines, which employ over 2,000 persons and are capable of turning out 100,000 tins a day.

Other manufactories. The limits of space forbid any, but a brief mention of other manufactories, of which a list is given below, the figures in brackets indicating the average daily number of operatives employed in 1911. There are four bone mills, viz., the Bengal (120), Ganges Valley (51) and Standard (59) Mills and the factory of the Agri-Phosphates Co. (95) at Ultādanga. At Cossipur there are a large sugar refinery, called the Cossipur Sugar Works (739) and a lac factory (420); lac is also manufactured in J. C. Galstaun's factory (153). Soap and candles are made in the North-West Soap Company's factory (183), silk at the Bengal Silk Mill at Ultadānga (181) and patent stone at the Indian Patent Stone Works (161). There are two ice factories, viz., the Calcutta and Linde Ice Factories, and the gas consumed in Calcutta is produced at the Oriental Gas Works (1,091). The latter have recently completed a very fine retort house, and have imported an expensive mechanical plant to charge and empty the retorts together with a plant for condensing, cooling and exhausting the gas, at a total cost of 4½ lakhs. This is said to be "the most up-to-date method of manufacturing gas to be seen in any country."† On the premises of the Linde Ice Company a new industry has lately been started, viz., the manufacture of oxygen gas for the purpose of carrying out repairs by the oxy-acetyline process, which is briefly as follows. Two jets impinge on the part to be welded, which immediately becomes incandescent from the intense heat generated; one jet is conveyed from the compressed oxygen, and the other from coal gas or an acetyline generator. The factory does not execute any repairs, but manufactures the gas for sale and is the only factory in India at which it can be obtained. It is made by eradicating the nitrogen from the atmosphere, and the oxygen is then pressed to 1,800 lbs. on the square inch and supplied in cast steel bottles.

---

* *Review of the Industrial Position and Prospects in Bengal in 1908.*

† Report of the Chief Inspector of Factories for 1911.

Two unsuccessful attempts have been made to manufacture glass in the district. The Pioneer Glass Manufacturing Co. worked at Titāgarh from 1890 to 1899, and the Indian Glass Co., and later the Bengal Glass Co., at Sodepur, from 1891 to 1902. Both had an expensive plant, but suffered from the difficulty of obtaining a supply of skilled labour which could carry on the trying work of glass-blowing in the climate of Bengal.

HAND INDUSTRIES.

Fishing.

Fishing is an industry of considerable importance, furnishing, according to the census figures, 63,000 persons with a means of livelihood. In addition to the numerous rivers, creeks and estuaries, the *bils* (of which some are large enough to be dignified with the name of lakes) are valuable fisheries. The most important are the Bālli Bils in the east, those to the south of Kāorāpukur and Thākurpukur, and the Salt Lake, from which Calcutta derives a large proportion of its daily fish supply. "In the Salt Lake," remarks Sir K. G. Gupta in his report on the Fisheries of Bengal (1908), "Calcutta has a valuable fishery of immense possibilities. At Dhāpa there are two municipal fisheries, both of which are let to lessees, who do not work them themselves, but simply sublet them to others at a considerable profit. The fisheries consist of several enclosures formed by throwing mud embankments round a shallow area along the side of a drainage channel communicating with the Salt Lake. Each of them is called a *bheri*, and water is let into it at flood time from the channel by a regulated wooden sluice, through which fry and small fish go but cannot come out again. The *bheris* act as nurseries, and fish of various kinds, chiefly prawns, small *bhektis*, *mugils* and *macrones*, are caught from time to time; but owing to the existence of annual leases the whole place is completely drained by February and all the fish caught, so that any large growth is not possible."

In addition to what may be called the natural sources of supply, the well-to-do systematically or occasionally stock their private tanks with fry obtained from the rivers. Such stocking is done as a commercial investment in the neighbourhood of Calcutta, and hundreds of people of both sexes find employment in gathering spawn, more especially in the Hooghly above Cossipur, and selling their catches to the tank stockers.

The list of fish caught for sale is a long one, but the following require mention. The most valuable fish caught in the estuaries and estuarine rivers are different kinds of *mugils* and *polynemus* and the well known *bhekti* (*Lates calcarifer*); the Hooghly from Diamond Harbour to Uluberia is noted for *tapsi* or mango fish (*Polynemus paradiseus*), which are caught in great numbers from April to June. *Hilsa* (*Clupea ilsha*) are also found in the estuaries and rivers in large quantities, and are netted in the rains, and for a few months after, when they ascend in shoals to breed. Of fresh-water fish the commonest are the *rui* (*Labeo rohita*), *kātla* (*Catla buchanani*), *mirgāl* (*Cirrhina mrigala*), *kālābāns* (*Labeo calbasu*), *koi* (*Anabas scandens*), *māgur* (*Clarius magur*) and *singhi* (*Saccobranchus fossilis*). The last three thrive in dirty stagnant water. In the smaller waterways within reach of Calcutta fishing is carried on throughout the year, but in the estuaries and larger channels of the Sundarbans it takes place only from October to March, after which a strong south-west wind sets in. The busiest season is from November to February, when parties of fishermen venture out to the sea-face.

Sugar making.

The manufacture of molasses, and to some extent also of sugar, is fairly extensive in the Bāduria thana of the Basīrhāt subdivision and at Gobardānga and Sukchar in the Bārāset subdivision. There were 22 small factories at work in 1910-11, the outturn being 8,000 maunds.

Gold and silver work.

A considerable business in gold and silver work is done at Bhawānipur by the firm of Messrs. Giris Chandra Dutt & Son. The articles are chiefly intended for European use, but have Indian ornamentation.

Other hand industries.

With these exceptions the hand industries of the district are of little importance. The weaving of coarse but durable cotton cloths on hand-looms still lingers on as a cottage handicraft. A finer cloth is turned out by the weavers of Tentra, a village that forms part of the Basīrhāt municipality. Mosquito curtains and the embroidered needle work called *chikan* are produced at Bārāset and its neighbourhood; the latter finds a ready sale among Europeans, and is exported to Australia and Europe. The manufacture of brass and bell-metal utensils is carried on at Basīrhāt, Tāki and Bāduria, chiefly for local consumption, and brass padlocks of good workmanship are turned out by a few families at Nātāgarh near Sodepur in the Barrackpore subdivision; locks are also made at Kādihāti in the Dum-Dum thana and Gopālpur in the Hāroa thana. There are some small tanneries at Tengra, where there is a colony of leather workers. Brushes are made at Khardah and Pānihāti, and musical instruments at Cossipur, Barnagore and Sinthi. Saltpetre is manufactured in refineries in the vicinity of Calcutta, and tobacco at Nawābganj and some places in the Bārāset thana. Mats are made at Itinda and Shaistānagar in the Basīrhāt subdivision, but the chief seat of the industry is the Falta thana of the Diamond Harbour subdivision, from which there is a considerable export to Calcutta. Palm-leaf braid for use in hats is made in this thana and also in the Diamond Harbour thana. Nut-crackers are made at Akarpur in Basīrhāt thana, and sacrificial knives at Gocha in the same thana and at Chitur in the Bāduria thana.

Trade. The principal articles imported by rail are coal from the Rāniganj and Jheria coal-fields, jute from East and North Bengal, and oilseeds from Bihār; the coal and jute go to the mills. Raw cotton is obtained by the mills from Berār and the Central Provinces, rice from Backergunge, Burdwān and Khulna, and paddy (unhusked rice) from Bīrbhūm and Bogra. Some gram and pulses are imported from Nadia and Jessore, and a little sugar comes from the latter district. Imported kerosine oil is sent up-country from Budge-Budge, the local rice goes to Calcutta, and gunny bags manufactured in the mills are despatched to Calcutta and upcountry. A considerable export takes place by road into Calcutta of animals, vegetables, etc., as well as of straw, bricks, bamboos and other local products and manufactures.

Fairs. A considerable amount of trade is carried on in the fairs or *melās* held periodically in the villages, of which a list is given below:—

NAME OF FAIR.	Place.	Approximate dates.	Name of thana and distance from headquarters.	Average attendance.
		SADAR SUBDIVISION.		
Gangā Pūja ...	Raipur ...	Last day of Paus	Budge-Budge, 20 miles	2,000
Ghāzi Sāheb's *melā*.	Bāsra	7th Asārh	Bāruipur, 18 "	7,000
Bhāngar Sāheb's *melā*.	Bhāngar Bazar	Month of Chaitra	Bhāngar, 21 "	8,000
		BARASET SUBDIVISION.		
Bārāset	Bārāset Court compound.	Srīpanchami day (2nd or 3rd week of January).	Bārāset, 18 miles	3,000
Khāspur	Khāspur ...	21st March or 1st week of February.	Ditto, 24 "	5,000
Kāzipāra	Kāzipāra ..	Paus Sankrānti or 2nd week of January.	Ditto, 20 "	3,000
Shāsan	Shāsan ...	5th Phālgun (2nd or 3rd week of February).	Ditto, 24 "	4,000
Budge-Budge ...	Budge-Budge	2nd Phālgun or 2nd week of February.	Ditto, 25 "	1,000
Baliāghāta ...	Baliāghāta ...	8th Chaitra or 3rd or 4th week of March.	Ditto, 28 "	6,000
Rahanda	Rahanda ...	12th Māgh or 4th week of January.	Ditto, 22 "	1,500
Jābilliāghāta ...	Jābilliāghāta	27th Māgh or 2nd week of February.	Ditto, 23 "	3,500
Shaimana	Shaimana ...	Just after Māghi Pūrnima or 1st or 2nd week of February.	Ditto, 22 "	1,500
Majlishpur ...	Majlishpur ...	26th Māgh or 2nd week of February.	Ditto, 25 "	1,500
Kotra	Kotra	2nd Māgh or 2nd or 3rd week of January.	Ditto, 25 "	1,500

NAME OF FAIR.	Place.	Approximate dates.	Name of thana and distance from head quarters.		Average attendance.
		BASIRHĀT SUBDIVISION—*concluded.*			
Khariberia ...	Khariberia ...	12th Phālgun or 3rd or 4th week of February.	Bārāsat	25 miles	250
Nebandha	Nebandha ...	Rās Pūrnima day or 2nd week of November	Ditto,	25 ,,	2,000
Kemia	Kemia ...	Ditto ditto ...	Ditto,	22 ,,	2,000
Bādur Sāheb's *melā*.	Bādurhāt	12th Māgh or last week of January.	Hābra,	35 ,,	10,000
Mantāj Mandal's *melā*.	Bauspol ...	5th Māgh or middle of January.	Do.,	35 ,,	8,000
Bara Pīr's *melā* ...	Jerat or Abalshidhi.	25th Phālgun or 2nd week of March.	Do.,	30 ,,	5,000
Ola Bibi's *melā* ...	Gaipur ...	1st Chaitra or 2nd week of March.	Do.,	48 ,,	2,000
Mānik Pīr's *melā*	Jangalpur ...	25th Phālgun or 2nd week of March.	Do.,	32 ,,	1,500
Ghāzi Pīr's *melā*	Bowgachi ...	20th Paus or 1st week of January.	Do.,	40 ,,	2,000
		BARRACKPORE SUBDIVISION.			
Bagjalapur Sāheb's *melā*.	Bagjala Ghojhādānga.	28th Māgh or 2nd week of February.	Dum-Dum,	25 miles	2,000 or 3,000.
Naihāti Mahākāli *melā*.	Naihāti ...	2nd or 3rd week of January ...	Naihāti,	18 ,,	100 to 200.
Kantālpāra Rāsh *melā*.	Kāntālpāra ...	Rāsh Pūrnima in Kārtik or 2nd week of November.	Do.,	18 ,,	4,000 to 10,000.
Bainchgācha Shāfarid's *mela*.	Bainchgācha	12th Māgh or 3rd week of January.	Do.,	20 ,,	200 to 250
Najdah Pāgla's *melā*.	Nagdaha ...	5th Phālgun or 2nd or 3rd week of February.	Do.,	20 ,,	50 or 60
Rāna Ekdil's *melā*	Rāna	22nd of Māgh or 1st week of February.	Do.,	25 ,,	300 to 500
Jerat	Jerat	25th Phālgun or 2nd week of March.	Do.,	28 ,,	4,000 to 5,000.
Tārāpukur Bara Pīrs *melā*.	Tārāpukur ...	14th or 15th January	Khardah	12 ,,	5,000
Khardah Phul Dol	Khardah ...	Baishākhi Pūrnima or 2nd week of May.	Do.,	15 ,,	2,000
Khardah Shām Sundar Jiu's Rās *melā*.	Ditto ...	Kārtik Pūrnima or 3rd week of November.	Do.,	15 ,,	3,000
Pinjrāpol Gopāstāmī *melā*.	Sodepur ...	3rd Agrahāyan or 19th November.	Do.,	14 ,,	3,000
		BASIRHAT SUBDIVISION.			
Gorāchānd's *melā*	Hāroa ...	24th February to 5th March ..	Hāroa,	38 miles	3,000
Mālancha *melā* ...	Mālancha ...	23rd November to 1st December.	Do.,	57 ,,	150
Mandar Sāheb's *melā*.	Mandar ...	23rd to 28th March	Hāroa,	36 ,,	200
Nebutalla Bārwāri *melā*.	Nebutolla ...	14th to 19th February	Do.,	38 ,,	300
Dhan Bibi's *melā*	Bolghātta ...	18th to 23rd February	Do.,	38 ,,	250
Kātia *melā* ...	Kātia	Baisākh (April-May)	Bāduria,	49 ,,	300
Tārāgunia Bāruni *melā*.	Tārāgunia ...	Chaitra (March-April) ...	Do.,	44 ,,	800
Atlia (Bura Pīrs *melā*).	Atlia	Last day of Agrahāyan or 15th December.	Do.,	48 ,,	1,000
Deara Junnia *melā*	Deara	Māgh (January-February) ...	Do.,	46 ,,	1,000
Sherpur *melā* ...	Sherpur ...	Ditto ditto	Do.,	36 ,,	500
Kutubzinda *melā*	Bāduria ...	Phālgun (February-March) ...	Do.,	44 ,,	200
Akidilka Mulla *melā*.	Mātābhanga...	Māgh (January-February) ...	Do.,	44 ,,	200
Shāhāchand *melā*	Andhār Mānik	Last day of Paus or 15th January.	Do.,	46 ,,	700
Madra Mulla (Kakurbar Pīr Sāheb's *melā*).	Madra ...	12th Māgh or 26th January ...	Do.,	44 ,,	150
Metia fair	Metia Hāt ...	5th March	Basīrhāt,	37 ,,	1,000
Begumpur fair ...	Begumpur ...	Ditto	Ditto,	35 ,,	500
Basīrhāt Bāruni *melā*.	Basīrhāt ...	23rd Chaitra or 6th April ...	Ditto,	45 ,,	1,000
Chaimalpur *melā*	Chaimalpur	28th t 30th March	Hāsanābād,	59 ,,	500

NAME OF FAIR.	Place.	Approximate dates.	Name of thana and distance from headquarters.	Average attendance.
		DIAMOND HARBOUR SUBDIVISION.		
Sripanchami Pūja	Hatugunj ...	February	Diamond Harbour, 34 miles.	5,000
Nanda *melā* ...	Barasi ...	April	Mathūrapur, 36 miles	5,000
Audhamuni *melā* ..	Krishnachandrapur.	March	Ditto, 34 ,,	2,000
Ganga Sāgar *melā*	Sāgar Island	January	Kulpi, 91 ,,	30,000 to 50,000.
Bishālakhi *melā* ...	Kāntabana ...	May	Do., 51 ,,	1,000
Kākdip *melā* ...	Kākdip ...	January	Do., 64 ,,	3,000
Keshabeswar *melā*	Mandira Bazar	April	Do., 31 ,,	6,000
Gota *melā*.. ...	Maiter Hāt ...	Do.	Do., 54 ,,	1,000
Tārāchānd *melā* ...	Lot No. 14 ...	February	Do., 62 ,,	3,000
Bāruni *melā* ...	Bishnupur ...	March	Magra Hāt 32 ,,	500
Do.	Multi	Do.	Do., 22 ,,	2,500
Paus Sankrānti *melā*.	Bishnupur ...	January	Do., 32 ,,	1,000

CHAPTER X.

MEANS OF COMMUNICATION.

NAVIGABLE CHANNELS. In a great portion of the district the waterways, such as estuaries, rivers and creeks, form the main, and often the only, means of communication. The most important navigable channels are included in the system known as the Circular and Eastern Canals or the Calcutta and Eastern Canals.

Circular and Eastern Canals. This is a system of natural channels, connected by a few artificial canals, in the districts of the 24-Parganas, Khulna, Farīdpur and Backergunge, which carry the produce of Eastern Bengal and the Brahmaputra Valley to Calcutta. They have a total length of 1,127 miles, of which 47 miles are artificial canals. The remainder are natural channels, mainly tidal creeks in the Sundarbans, which stretch eastwards from the Hooghly across the Ganges delta and afford means of intercommunication between the rivers and estuaries. The channels are under the supervision and control of Government, and tolls are charged on vessels when they enter the Circular Canal at Dhāpa lock, 5 miles east of the Hooghly. Dhāpa, which is the western terminus of the system, is connected with Calcutta by the Baliaghāta Canal and with the northern suburb of Chitpur by the New Cut and Circular Canals. To the east the objective of the system is Barisāl, the headquarters of the great rice-growing district of Backergunge, situated 187 miles from Calcutta.

This is one of the most important systems of river canals in the world, judging by the volume of its traffic, which averages a million tons per annum, valued at nearly four million sterling. The situation of Calcutta makes it the natural outlet for the Ganges valley, and its position has been enormously strengthened by the construction of railways; but other measures were necessary to enable it to tap the trade of the Brahmaputra Valley and to focus the traffic of the eastern districts. The intermediate country is a maze of tidal creeks, for the most part running north and south, but connected here and there by cross-channels, wide near the sea-face but narrow and tortuous further inland. These inland channels are constantly shifting as the deposit of silt raises their beds, while, on the other hand, the great estuaries near the sea-face are not navigable by country boats from June to October, owing to the strong sea-breezes which prevail during the south-west monsoon. This system of channels was devised, therefore, in order to allow country boats to pass from the eastern districts to Calcutta by a direct inland route, and the problem has been to keep the natural cross-channels clear of silt, and to connect them with each other and with Calcutta by a system of artificial canals.

When British rule began, boats from the eastern districts could only approach Calcutta by a route close to the sea-face, which brought them into the Hooghly by the Bārātala creek, 70 miles below Calcutta ; this route was not only circuitous, but also impracticable for country boats during the monsoon. The pioneer of the present system was Major Tolly, who in 1777 canalized an old bed of the Ganges from its confluence with the Hooghly at Hastings, a little to the south of Fort William, south-eastwards to Garia, a distance of 8 miles. From this point the canal (which was called Tolly's Nullah after him) was carried east to meet the Bidyādhari river at Samukpota, and thus gave access to an inner route which leads eastwards from Canning. In 1810, a further step was taken to improve the approaches to Calcutta, an old channel through the Salt Water Lakes, east of the city, being improved and led westwards by what is now known as the Baliaghāta Canal in the neighbourhood of Sealdah. Between 1826 and 1831 a new route was opened between Calcutta and the Jamuna river, following the same direct easterly course as the present Bhāngar Canal, the object being to relieve the pressure on Tolly's Nullah ; a number of tidal channels were utilized and connected by six cuts to form a continuous eastern route.

The next step was to make the Circular Canal from Chitpur, parallel with the Circular Road, to meet the old Eastern Canal at Baliaghāta ; this was completed in 1831. The canals were still choked by the ever increasing volume of traffic, and in order to relieve them, the New Cut was opened in 1859 : this leads south east from Ultadānga, on the Circular Canal, 3 miles east of Chitpur, to Dhāpa on the Baliaghāta Canal. After this, the Bhāngar channel was canalized in 1899 for a length of 15 miles, thus completing the inner channel which had been commenced in 1831. Next, a channel was made from Ultadānga to Bāmanghāta, in order to save the boats from the eastern districts from having to come from Bāmanghāta to Dhāpa through the congested Salt Lake channel ; on entering the canal at Kulti they can proceed to Calcutta without further lockage. The new canal, which was completed in 1910-11, takes off from the New Cut canal, near where the railway crosses it and alongside the Arathoon Mills, and joins the Bāmanghāta-Kulti canal. It passes round the Northern Salt Lake, and facilitates drainage as well as navigation, for it receives the drainage of the country near Dum-Dum, and carries it away by means of a large sluice.

This system of navigable channels was devised and has steadily been developed for boat traffic, and there are not the same facilities for steamers. The whole of the steamer traffic from the eastern district has still to find its way to Calcutta by a long circuitous route through the Sundarbans and round by Sāgar Island. Proposals for making a steamer canal between Calcutta and Eastern Bengal have been brought forward from time to time, and about 35 years ago the Bengal Government was prepared to construct such a canal; but the steamer companies were not in favour of the scheme. At that time, the water-borne traffic to Calcutta was seven times as great as that carried by the Eastern Bengal State Railway, and the inland steamer companies could dictate their own terms for the carriage of goods. They preferred the route through the Sundarbans, because, though long, it was free from tolls, to a direct canal, for the use of which considerable tolls would be charged. With the development of railways in Bengal and Assam, the agents of the steamer companies have come to realize that the circuitous route through the Sundarbans is a great disadvantage, and the question of providing a short direct route has been revived. In this connection a proposal has been made for the canalization of Tolly's Nullah, the effect of which would be to shorten the existing route by 186 miles, and bring the river steamers running to and from Eastern Bengal and Assam into closer touch with the trade centres of Calcutta. The present position is that there is, for about three months in the year, fairly efficient water communication with the eastern districts as far as Sāmukpota, at the junction of Tolly's Nullah with the Bidyādhari river, but beyond that place steamers are blocked from access to Calcutta. They cannot pass through the Dhāpa lock into the Calcutta canals, and they are debarred from entry into Tolly's Nullah, as it is only a creek which runs almost dry at low tide. They are consequently shut off from direct connection with the two great markets of Calcutta, viz., Chitpur on the north and Kidderpore on the south. In order to obtain access to the Hooghly river, and so to them, they have to follow, instead of the route *viâ* Sāmukpota, a circuitous course through the Sundarbans and Channel Creek, which adds about 200 miles to the length of every voyage.

In order to remedy this state of affairs, Mr. O. C. Lees of the Public Works Department proposed in 1902 that Tolly's Nullah should be canalized so as to provide a channel, from the Bidyādhari river to the Hooghly, capable of being used by the large inland steamers and flats employed in the trade with Assam and Eastern Bengal. In 1904 the proposal was placed before the Government of India, the total cost (including dredging plant) being estimated at Rs. 1,19,79,320. That Government, while agreeing that the improvement of Tolly's Nullah was desirable in order to provide direct access to Calcutta, held that the provision of funds for so costly an undertaking prevented its acceptance. The discussion was subsequently reopened in 1906, and since that time various proposals and revised estimates have been made.

Grand Trunk Canal Project.

A larger and more comprehensive scheme was put forward by Mr. O. C. Lees in 1913, the total estimated cost of which is Rs. 2,10,15,000. This contemplates the construction of a Grand Trunk Canal with the following alignment.

Starting from Badartala on the Hooghly, just beyond port limits, it takes a southerly course for the first two miles and, after crossing the Budge-Budge Railway, continues in a south-easterly direction to Putiāri. Taking an easy curve to the east, it then crosses Tolly's Nullah at Bānsdhani ferry and, continuing in an easterly direction, enters and leaves the Bidyādhari river rather more than a mile above Sāmukpota and terminates on the right bank of the Kulti river about a mile below the entrance to the Bhāngar canal, at the place where the river bifurcates. The southern branch, which formed the old channel of the Kulti river, is now practically closed, but the other branch, which runs in an easterly direction and now forms part of the "Outer Boat Route," is in good order, has ample depth at low water and will require comparatively little to be done in easing sharp bends and widening narrow reaches to convert it into a good navigable channel for steamer traffic. From the Kulti end of the canal, the "Outer Boat Route" will be followed. The improvements in that route required to enable it to satisfy the requirements of steamer, as well as boat, traffic present no difficulties that cannot be surmounted by the employment of suitable dredging plant. The most costly part of the project will be the new canal to connect the Hooghly river with the Kulti river and the improvement of the existing channel of Tolly's Nullah from its junction with the new canal to its entrance at Hastings.

As explained in Chapter I the Bidyādhari river has silted up seriously, and it is proposed to canalize both that river and the Piāli river. On this point Mr. Lees writes:—"Regarded from the navigation point of view, the canalization of the Bidyādhari and Piāli rivers will be very advantageous. In the old days, when the lower reaches of the Bidyādhari presented less dangers to navigation than they do at present and when Tolly's Nullah provided a good and easy approach to Calcutta, there was a large country-boat traffic along this route. The firewood traffic, especially, was considerable, and it has been killed partly by the dangerous navigation of the Bidyādhari and partly by the closure of Tolly's Nullah. When, however, these large country-boats are able to get into the Piāli river from the Mātla and are provided with a safe and good canalized route to Calcutta *viâ* the old channel of Tolly's Nullah and the connecting channel, there will be an immediate resuscitation of the firewood traffic and a great development of general boat traffic. The enlargement of the upper reaches of the Bidyādhari will, moreover, restore direct communication between the Sundarbans and the Chitpur canal."

From the Kulti terminus of the Grand Trunk Canal the new steamer route will follow the line of navigation known as the "Outer Boat Route." The channels for the most part are in good order and quite suitable for inland steamer traffic, but improvements are required along portions of the route. As far as the junction of the Chunkuri Khāl with the Passur river, the main route will be common to all the traffic eastwards from Kulti. The channels on the main route, which require to be enlarged or improved to render them suitable for steamer as well as boat traffic, are all within the first 100 miles between Kulti and the Passur junction. Following the route eastwards from Kulti, the first channel requiring improvement is the Kulti river itself. It is in a very healthy condition and the depth is ample, but in places the channel is too narrow and there are several very abrupt bends. The estimate provides for increasing the width of the channel to 250 feet at a depth of 10 feet below low-tide level and easing the sharp bends. The length to be improved is about 8 miles.

The Inner Boat Route.

There are three alternative routes to Barisāl known as the Inner Boat Route, the Outer Boat Route, and the Steamer Route. The Inner Boat Route, which is used by small country boats and launches, passes from Chitpur by the Circular Canal and the Lake Channel, or from Kidderpore by Tolly's Nullah, to Bāmanghāta, where it enters the Bhāngar Canal. Thence it proceeds by connected waterways to Hāsanābād on the Ichāmati or Jamuna river and down that river to Basantpur in the Khulna district.

The Inner Boat Route is usually followed, by preference, by small country boats. It is too narrow, however, for the great Sundarbans wood boats, some of which carry 1,000 maunds or more, and for other large country boats, and these are compelled to follow the Outer Boat Route. This is also used by small cargo steamers, for nothing larger than a steam launch can traverse the Inner Route. The Outer Route starts from Sāmukpota, 20 miles south-east of Calcutta, which is reached from that city either by the Circular Canal or by Tolly's Nullah Thence it proceeds south-east down the Bidyādhari river to Canning, from which place it crosses by several channels to the Kālindi river, and follows that river up to Basantpur, a distance of 54 miles from Sāmukpota. **The Outer Boat Route.**

The Steamer Route is used by the large steamers and flats of the India General and Rivers Steam Navigation Companies. It proceeds from Calcutta down the Hooghly to Mud Point, where it passes half-way down the Bārātala river, or Channel Creek, between Sāgar Island and the mainland. Thence it passes by a cross channel into the Sabtarmukhi and across it into the Jāmira. Other cross channels take it across the Mātla, Guāsuba, Hāriābhānga and Kālindi, successively, into the Khulna district, from which it works its way eastward to Barisāl. **The Steamer Route.**

The following canals are under the charge of the Executive Engineer, Circular and Eastern Canals Division. The term 'canals' is applied to natural channels as well as to artificial canals; as a matter of fact the only channels actually canalized by having locks at each end are the Circular Canal (including the Baliaghāta Canal) and the Bhāngar Canal:— **CANALS.**

Name.	Length in miles.
Circular Canal (including Baliaghāta Khāl) ...	$5\frac{1}{2}$
New Cut Canal	4
Lake Channel (from Dhāpa to Bāmanghāta) ...	$5\frac{1}{2}$
Eastern Canals (from Bāmanghāta to Basantpur)	42
Outer Boat Route	4
Ditto (from Sāmukpota *viâ* Mātla river to Basantpur)	54
Tolly's Nullah	17
Kaorāpukur Khāl	20
Sundarbans Steamer Route	172

The following is a brief description of some of the more important canals:—

Tolly's Nullah, 17 miles in length, extends from Kidderpore to Tārdaha, and so connects the Hooghly with the Bidyādhari **Tolly's Nullah.**

river. It was originally a silted up channel of the Ganges, and was excavated in 1776 by Major W. Tolly as a private venture, under a temporary grant of land and of the right to levy canal tolls. The nullah was opened to navigation in 1777; in 1790 we find an advertisement in the *Calcutta Gazette* (of 18th March) offering for sale the remaining term of the lease granted by Mary Anna Maria Tolly to John Wilkinson, from which it appears that the tolls were leased for 10 years at a time. The canal was taken over by Government in 1804; an official announcement in the Gazette states that the tolls on boats and goods passing through the Tolly's Nullah, formerly levied for the benefit of the late Mrs. Tolly, are henceforth to be collected for Government under the superintendence of the Collector of the 24-Parganas. When first excavated, the canal was of insignificant dimensions, but it has since been widened, and is now a much frequented route. It is, however, very liable to silting, especially near Tollygunge, where the tides meet, and has to be constantly cleared in order to keep it navigable. Even in the height of the rains a steam launch can pass Tollygunge only when the tide is half full or higher.

Kāorāpukur Canal.

The Kāorāpukur Canal or Khāl, 20 miles long, branches off from Tolly's Nullah, a few miles south of Calcutta, and runs southward to Magrā Hāt.

Circular, Baliaghāta and New Cut Canals.

The Circular Canal extends from Chitpur lock to the lock at Dhāpa, a distance of 5½ miles, and forms the north-eastern boundary of Calcutta, which it separates from Māniktala. The section which forms the southern boundary of the latter municipality is also known as the Baliaghāta Canal. A branch canal from Ultadānga (on the Circular Canal) to Dhāpa is known as the New Cut Canal. Māniktala is entirely surrounded by these three canals, viz., the Circular Canal on the west, the New Cut Canal on the north and east and the Baliaghāta Canal on the south.

Lake Channel.

The Lake Channel is a tidal river, 5½ miles long, from Dhāpa to Bāmanghāta, which has silted up considerably and can only be maintained for the passage of boats by periodical silt clearance.

Bhāngar Canal.

The Bhāngar Canal or Khāl, 15 miles long, extends from Bāmanghāta on the Bidyādhari river to Kulti lock on the channel called the Kulti Gāng. Its construction was commenced in 1896 and completed in 1897.

Kristopur Canal.

The Kristopur Canal, 10 miles long, connects the New Cut Canal with the Bhāngar Khāl, and saves boats from the eastern districts from having to pass through the Dhāpa and Bāmanghāta locks.

Besides the above there are three small channels kept up by the District Board. They are navigable only by *dongās* or dug-outs, and by them only for about eight months in the year, being almost dry from January to May. They are—(1) Serākol to Jabaral, along the Diamond Harbour Road, 11¾ miles long There are five bridges over the *khāl* and a sluice at Jabaral to allow of the ingress and egress of the tide. (2) Sāpgāchi to Chaubāga (near Dhāpa), 2 miles long. This *khāl* was excavated in 1882 at the cost of the Government Khās Mahāl Department, and also serves as a drainage channel for the benefit of the Khās Mahāl lands at Dhāpa. (3) Magrā Hāt to Jaynagar, 6½ miles long. **District Board channels.**

There are 53 ferries under the control of the District Board, the most important being the Uttarbhāg ferry across the Piāli river, the Hājipur ferry across the creek at Diamond Harbour, and the Budge-Budge and Charāmādāri ferries across the Hooghly. The boats which serve these ferries are mostly large strongly built country boats of 50 to 100 maunds burthen, which can carry 25 to 50 passengers. Other ferries on the Hooghly river belong either to the Government, the Port Commissioners, or the riparian municipalities. **Ferries.**

In addition to the Sundarbans steamer services, passenger steamers ply on the Hooghly from Calcutta to Naihāti, from Calcutta to Kākdwīp on the main land opposite Sāgar Island, and from Diamond Harbour to Tengra. Steamers also ply on the Ichāmati and Jamuna rivers from Tāki to Gobardānga (42 miles). **Steamer services.**

The greater part of the inland traffic is carried by country boats, the largest of which can carry as much as 150 tons. They are generally very broad in the beam and of light draught, and have a large square sail. Against the wind they are rowed, or poled if the water is shallow; but, as a rule, they travel with the tides, going up one river with the flood and down another with the ebb, and anchoring when the tide is against them. The local boat is called a *pānsi*; it is a broad-bottomed boat, with a thatched cabin and a deck made of bamboos or planking. **Boats.**

The *dingi* is generally used on the larger rivers for passenger traffic. This is 25 to 30 feet long and has a breadth of about 4 feet. It has a bamboo mast, and an arched roof of matting in the middle affords protection against the weather. Its average burthen is 12 to 15 maunds, and it is managed by two men, one in the bows and the other (who steers it) in the stern. Traffic on the smaller streams and across the swamps is carried by light dug-outs, which are poled in shallow water and paddled in the deeper channels. One kind, called the *dongā* or *sālti*, is made of the hollowed-out trunk of a large tree or *sāl* tree (whence the name), and is some 20 feet or more long, 2 feet broad at the gunwale and about 2½ feet broad at the bottom, which is flat: sometimes a small matting roof is rigged up in the centre. It is propelled by two men, one at either end, and a European has to travel in it sitting down; it would probably upset if he tried to stand up. The *telo dongā* is a much smaller dug-out, made of the trunk of a palm tree. One man paddles or poles it, and it can carry only one passenger seated precariously at the other end—a true "siege perilous," for some are barely 9 or 10 feet long by one foot in breadth and depth.

RAILWAYS

Eastern Bengal State Railway.

The district is served by the Eastern Bengal State Railway, which has its terminus at Sealdah, just within the boundaries of Calcutta. The workshops and headquarters of the locomotive department are situated at Kānchrāpāra on the northern boundary of the 24-Parganas. The line originally belonged to a private company, and was taken over by Government in 1884; it was transferred from Provincial to Imperial control in 1892. The line in this and other districts south of the Ganges is on the standard gauge of 5 feet 6 inches, and includes the following sections:—

(1) Eastern Section from Sealdah to the northern boundary of the 24-Parganas, the last station within district limits being Kānchrāpāra, 28 miles from Sealdah.

(2) Southern Section, with three branches running to Diamond Harbour, Budge-Budge and Canning, viz. (*a*) main line from Baliaghāta to Diamond Harbour, 37 miles; (*b*) Budge-Budge branch from Ballygunge Junction to Budge-Budge, 13 miles; (*c*) Canning branch from Sonārpur Junction to Canning, 18 miles.

(3) Central Section from Dum-Dum Junction to Gobardānga, 31 miles, with a branch from Dum-Dum to Pātipukur, 4 miles, and another branch from Bārāsāt Junction to Baliaghāta Bridge, 9 miles.

The Eastern Section serves the north-west of the district lying within thanas Dum-Dum, Barnagore, Barrackpore, Nawābganj, Khardah and Naihāti. Work was started in 1859, and the line was opened as far as Rānāghāt in the Nadia district (45 miles) in 1862. It was doubled as far as Naihāti in 1886, up to Rānāghāt in 1892, and to Porādaha in the Nadia district in 1897. It is proposed to quadruple the line as far as Naihāti, and work is in progress. Recently, in order to provide direct access to the Kidderpore Docks, and to relieve the main line of a large number of goods trains, the line from Dum-Dum Junction to Kānkurgāchi (a mile north of Sealdah) 3 miles long, has been quadrupled, and a new chord line from Kānkurgāchi to Gobra on the Southern Section, 2 miles long, was opened in 1907.

There are two branch lines to Chitpur and Naihāti in connection with this section. The Chitpur branch, which carries goods traffic only, leaves the main line near Belgāchia and joins the Port Commissioners' Railway at Chitpur. Its length is two miles, one half being on a heavy embankment; it is carried over the Dum-Dum road on a girder bridge. This branch was opened in 1873. From Naihāti a short branch line runs across the Jubilee Bridge on the Hooghly to Hooghly, on the main line of the East Indian Railway, a distance of 4 miles. This branch belongs to the East Indian Railway and was opened in 1887.

Southern Section.—The line running to Canning through Ballygunge, Garia, Sonārpur and Chāmpāhāti was originally known as the Calcutta and South-Eastern Railway, and was costructed by a private company, under a Government guarantee, for the purpose of receiving the traffic expected to accrue from the opening of Port Canning as an auxiliary harbour to Calcutta. The line was opened up to Chāmpāhāti in 1862, and was completed and opened throughout in 1863; but the failure of Port Canning involved the railway in its ruin, and in 1868 the line was taken over by Government. This branch serves thanas Ballygunge, Tollygunge, Sonārpur, Bāruipur and Port Canning. It is single, except from Sealdah to Ballygunge, this part being doubled in 1888. Sufficient land, however, was taken up for a double line, and portions of an embankment for a double line have been constructed. The masonry for the bridges has also been built for a double line, but the girders have been laid only for a single line, except on the bridges across Tolly's Nullah at Garia and the Piāli river, which are adapted for a double line. The terminus of the line was originally at Baliaghāta; since 1885, when it was joined on to the main line, it has been at Sealdah. A double line from Ballygunge to Bāruipur has recently been sanctioned and is under construction.

The Diamond Harbour line leaves the Port Canning line at Sonārpur and runs in a south-westerly direction to Diamond Harbour. It serves thanas Sonārpur, Bāruipur, Magrā Hāt, and portions of Jaynagar and Diamond Harbour. The construction of this line was commenced in 1880, and it was opened as far as Magrā Hāt in 1882, and completed up to Diamond Harbour in 1883. Except for the Usti Khāl bridge, which has a single span of 100 feet, there are no engineering works of importance on the line. A small branch, opened in 1883, runs from

Diamond Harbour southwards for half a mile to Hāra, on the Hooghly.

The Budge-Budge line serves the Budge-Budge thana and a portion of Tollygunge. It was sanctioned in 1886 as a line from Ballygunge to the Kidderpore Docks, which was to be constructed by the Port Commissioners; a tramway from Tolly's Nullah to Santoshpur, which belonged to the Port Commissioners, was taken over and incorporated in the line. Sanction to an extension to Budge-Budge was given in 1888, and the whole line as far as Budge-Budge was opened in 1890. The line is double as far as Bracebridge Junction, and single from there to Budge-Budge. The only engineering work of importance is the bridge over Tolly's Nullah, the main span of which is 115 feet 10 inches. A small branch, 2½ miles long, runs from Bracebridge Junction to Bracebridge Hall, on the Hooghly, and is used for carrying coal to the British India Company's steamers which coal there.

The Central Section follows a north-easterly course through Bārāset and Hābra, and serves thanas Dum-Dum, Bārāset and Hābra. The line in this district forms part of a line running from Sealdah to Khulna, with branches from Bangaon to Rānāghāt (in Nadia) and from Dum-Dum to Pāttipukur in the 24-Parganas. It was originally the property of the Bengal Central Railway Company, and was purchased and amalgamated with the Eastern Bengal State Railway in 1905, when the term of the contract with the Company expired. Construction was begun by the Company in 1881, and the line, which is single throughout, was opened up to Gobardānga in 1883.

Port Commissioners' Railway. The Port Commissioners' Railway connects the Eastern Bengal State Railway with the docks, and runs from Cossipur (near the Government Shell Foundry) to the Kidderpore Docks. It carries goods traffic only and serves the godowns, warehouses and mills along the Hooghly, the jetties and the docks. It was opened for traffic in sections, viz., from Bāghbazar to Mirbohur Ghāt in 1875, from Bāghbazar to Cossipur and from Mirbohur Ghāt to the jetties in 1878, from the jetties to Chāndpāl Ghāt in 1880, and from Chāndpāl Ghāt to the Kidderpore Docks in 1891. The only engineering works of importance are the bridges over the Circular Canal at Chitpur and over Tolly's Nullah at Hastings. The platforms carrying the rails on these bridges can be raised and lowered so as to allow boats to pass at certain states of the tides.

Light Railways. There is a light railway known as the Bārāset-Basirhāt Light Railway, which belongs to a limited liability company, and runs from Pāttipukur to Chingrihāta (Hāsanābād), a distance of

43 miles. It is on the 2 feet 6 inch gauge and is laid along the District Board road. It was constructed, and is managed for the company, by Messrs. Martin and Company, under an agreemen t with the District Board by which the latter guarantees interest at the rate of 4 per cent. up to a maximum of Rs. 38,000 per annum. The length from Bārāset to Basīrhāt (26 miles) was opened to traffic in 1905, an extension to Tāki and Hāsanābād (8 miles) in 1909, and the line from Baliaghāta Bridge to Pāttipukur (on the Dum-Dum road 2 miles north-east of Calcutta), which is 16 miles long, in 1910.

The suburbs of Calcutta are served by the Calcutta Electric Tramway, which has been brought out to Alipur, Behāla, Belgāchia and Tollygunge. **Tramways.**

Acording to the returns for 1911-12, the roads of the district have a length of 1,690 miles, viz., (1) Provincia roads, metalled, bridged and drained throughout, 88 miles; (2) District Board roads, metalled, bridged and drained throughout, 179 miles; (3) District Board roads, unmetalled but bridged and drained throughout, 292 miles; (4) District Board roads, unmetalled and only partially bridged and drained, 43 miles; (5) District Board roads, banked and surfaced with *murram* (gravel) or similar material but not drained, 46 miles; and (6) village roads, 1,042 miles. Over one-half of the roads are village roads, which are rough tracks passable only in fair weather. Only the two first classes, with an aggregate length of 264 miles, can be regarded as passable for wheeled traffic throughout the year. **Roads.**

The principal roads are as follows:—The Grand Trunk Road runs north from Calcutta along the river bank to Falta (14 miles), where the Hooghly is crossed by a ferry, and the road passes to the west bank of the river. The Ghoshpāra, or Plassey, road continues north along the east bank to the boundary of the district. The Jessore road passes through Dum-Dum and Bārāset and maintains a north-easterly direction to the district boundary. The chief roads south of Calcutta are the Diamond Harbour Road, the Orissa Trunk Road, which crosses the Hooghly by a ferry at Achipur, and the Vishnupur road, which runs due south, through Bāruipur, for 29 miles. The chief east and west cross-roads are the Tāki road from Bārāset to Basīrhāt, and thence to Tāki and Hāsanābād (along which the light railway has been laid), and the continuation of this road westwards to Barrackpore. The Provincial roads which are maintained by the Public Works Department are shown below :—

Name.	From To		M.	F.
1. Grand Trunk Road ...	From Tāla to Falta Ghāt	...	14	1
2. Calcutta-Jessore Road ...	From Calcutta to Chongda	...	28	6

Name.	From To	F.	M.
3. Cossipur-Dum-Dum Road	From Cossipur Ghāt to Dum-Dum	2	0
4. Diamond Harbour Road	From Durgāpur to Diamond Harbour	25	3
5. Orissa Trunk Road (or Budge-Budge Road)	From 3rd milestone of Diamond Harbour Road to Achipur on river Hooghly ...	13	1
6. Barrackpore Park Road ..	From 14th mile of Grand Trunk Road at Titāgarh to south west gate of Government Park at Barrackpore ...	0	5
7. Akra	Metiābruz Road	4	0

The Grand Trunk, Jessore and Orissa Trunk Roads are long roads extending through other districts, and only the lengths lying within the 24-Parganas are shown.

CHAPTER XI.

LAND REVENUE ADMINISTRATION.

DEMAND OF LAND REVENUE. The total demand of land revenue, according to the returns for 1911-12, is Rs 18,01,925, or approximately 18 lakhs. The gross rental of the district, as ascertained from the road and public works cess valuation, is 72½ lakhs, or four times the demand of land revenue. As the normal area under cultivation is 1,075,800 acres, the incidence of the gross valuation per acre amounts to Rs. 6-12, of which the share claimed by the State only represents one-fourth.

ESTATES. The demand is distributed among the different classes of estates as follows. There are 1,697 permanently-settled estates with a demand of Rs. 12,60,887, and 303 temporarily-settled estates with a demand of Rs. 1,55,980, while 33 estates, paying Rs. 3,85,058, are held directly by Government. Many of the landlords of the Bāraset and Basīrhāt subdivisions ordinarily reside at Calcutta, and rarely come into personal contact with their tenants, their estates being left to the management of *naibs*, or local agents.

There are 3,114 revenue-paying estates, 287 revenue-free estates, 12,869 rent-free lands and 191 *hāts* and fairs assessed to road and public works cesses, the aggregate current demand being Rs. 3,43,167. The recorded shareholders of these 16,461 estates are 25,552 in number There are also 53,331 tenures assessed to cesses with 87,931 shareholders. The gross rental of the district has increased by 28⅔ lakhs since road cess was first assessed under Act X of 1871.

Wards' Estates.

In January 1913, seven estates, as shown in the margin, were under the management of the Court of Wards. Only a portion of the Tagore estate lies in the 24-Parganas, and 74,691 acres are included in Eastern Bengal districts.

Name of estate or number of group of small estates.	Area in acres.	Rent-roll.
		Rs.
Janbazar	49,600	1,45,563
Bawāli	7,680	40,807
Gobardānga	38,000	58,234
Gopāl Chandra Mukharji and another.	21,667	54,991
Estate of Rājā Sir Saurīndra Mohan Tagore.	155,286	1,17,317
Pānihāti I	33,142	35,166
Do. II		11,511
Do. III		19,423
Do. IV		16,891

Government Estates.

The district contains a large area managed direct by Government, from which indeed more than a sixth of the land revenue is derived. The whole of the Sundarbans is so dealt with, and in the northern portion of the district there are several Government estates with an interesting history. Among these may be mentioned the Barnagore estate, which was acquired by treaty from the Dutch in 1795, and the Sāhibān Bāgīcha, which is an estate formed (in 1790) of garden houses occupied by Europeans. Another Government estate of an unusual character is the Orphanganj market, which is a much frequented market situated at Kidderpore.

The most important of the Government estates is Panchānnagrām, which is situated in the suburbs of Calcutta. It is so named from the fifty-five villages which the estate originally comprised; these, according to Holwell, were, in 1757, "taken from the Twenty-four Parganas adjoining to Calcutta in order to extend its bounds." The estate has an area of 26 square miles, and the lands comprised in it lie round Calcutta on the south and north-east beginning at the Government telegraph yard on Tolly's Nullah to the south, running up to Dum-Dum on the east and bounded on the north by the Barnagore estate. The land revenue demand is Rs 1,07,000, derived from nearly 15,000 holdings. A portion of the estate pays a fixed rate of Rs. 3 per *bigha*, *i.e.*, about Rs. 9 an acre, and in the remainder, rates fluctuate according to the position and advantages of the land.

SETTLEMENT OF LAND IN THE SUNDARBANS.

The first comprehensive set of rules for the grant of leases of land in the Sundarbans was issued in 1853. Leases were to be sold to the highest bidder when there was more than one applicant; and it was stipulated that one-fourth of the area should be permanently exempted from assessment, so as to allow for sites for houses, water channels, embankments, etc. The remaining three-fourths was to be free from assessment for twenty years, and to be assessed thereafter at progressive rates, which were decidedly light, for only half an anna a *bigha* was charged at first and the maximum rate (from the 51st to the 99th year) was only two annas a year. After the 99th year, the grant was to be liable to survey and reassessment at such rate as Government might think fit, the grantee, his heirs and assigns having such rights as to taking settlement as are generally applicable to temporarily settled estates. The rules further provided that one-eighth of the grant should be cleared and rendered fit for cultivation by the end of the fifth year, one-fourth by the end of the tenth year, one-half by the end of the twentieth year, and three-fourths by the end of the thirtieth year. Failure to carry out this clearance entailed forfeiture of the grant. The number of leases granted under these rules in the 24-Parganas and Khulna Sundarbans was 131; the rent payable is Rs. 1,33,447, which will eventually rise to Rs. 1,35,802. **Rules of 1853.**

In 1861 a new policy was introduced by Lord Canning, viz., that of disposing of waste lands in fee-simple, or, in other words, selling them revenue-free. Three main principles were laid down in Lord Canning's minute on the subject. (1) That "in any case of application for such lands they shall be granted in perpetuity as a heritable and transferable property, subject to no enhancement of land revenue assessment." (2) That "all prospective land revenue will be redeemable at the grantee's option by a payment in full when the grant is made; or, at the grantee's option, a sum may be paid as earnest, at the rate of 10 per cent., leaving the unpaid portion of the price of the grant, which will then be under hypothecation, until the price is paid in full." (3) That "there shall be no condition obliging the grantee to cultivate or clear any specific portion within any specific time." The minimum price of the fee-simple was fixed at Rs. 2-8 an acre, so that by paying 10 per cent. of this, or four annas an acre, a title was obtained. Subsequently the Secretary of State issued a despatch, which laid down that grants should be surveyed before sale, and that all sales should be by auction to the highest bidders above a fixed upset price. The fee-simple rules superseded those of 1853 and remained in force until 1879, when they were withdrawn and a fresh set of rules was issued. **Fee-simple rules.**

The rules of 1879 provided for two kinds of grants, viz., (1) blocks of land not exceeding 200 acres, leased to small settlers, and (2) blocks of 200 acres or more, leased to large capitalists who were prepared to spend time and money in **Rules of 1879.**

developing them. These grants are known as "lots," their holders being called *lotdārs*. Between 1879 and 1904, when the rules were suspended, grants or leases were made of 1,223 square miles, out of an available area of 2,301 square miles.

Small capitalist rules.

The small settlers were guaranteed a lease for 30 years if the lands were brought under cultivation within two years. They were allowed a rent-free term of two years, after which progressive rates of rent were fixed on the cultivated area, with reference to the rates paid in the neighbourhood, by raiyats to landlords, for similar lands. If available, an area of unclaimed land equal to the cultivated area was included in the lease, in addition to which the lessee could bring under cultivation any quantity of land adjoining his holding which was *bona fide* uncultivated. The holding was liable to measurement every five years, and all cultivated land in excess of the area originally assessed could be assessed at the same rate. After 30 years renewed leases could be given for 30 years periods, the rates of assessment being adjusted at each renewal with reference to the rates prevailing in the neighbourhood. No charge was made for wood and timber on the grant, nor for any cut or burnt in making clearances or used on the land, but a duty was to be levied on any exported for sale.

Large capitalist rules.

Under the rules for large capitalists the maximum area of grants was restricted to 5,000 *bighas*, the minimum being 200 *bighas*. The term of the original lease was fixed at 40 years, and after its expiry resettlements were to be made for periods of 30 years, maximum rates being laid down for each resettlement. One-fourth of the area was exempted from assessment in perpetuity, and the remainder was held free of assessment for ten years. It was stipulated that one-eighth of the entire area must be rendered fit for cultivation by the end of the fifth year, and this condition was enforced either by forfeiture of the lease or by the issue of a fresh lease omitting the remainder of the rent-free period and requiring payment of rent at enhanced rates. The rules also provided for gradually increasing rates of rent after the expiration of the rent-free period and for varying rates within different tracts according to the rent-paying capabilities of the land. The limits within which lands might be leased were to be fixed in consultation with the Forest Department, and rights of way and water and other easements were reserved. The right of using all navigable streams, and the use of a tow-path not less than 25 feet wide on each side of every such stream, were also reserved to the public. No charge was made for timber on the land at the time it was leased, nor for any cut or burnt to effect clearances or used on the land, but a duty was levied on any exported for sale. Leases were sold at an upset price of Re. 1 an acre when there was only one applicant, and to the highest bidder when there was more than one. Altogether, 188 leases have been granted under the "large capitalist rules" the rent payable being Rs. 70,329, which will rise eventually to Rs. 2,35,111.

Experience showed that this system was liable to abuse, and that the actual cultivators were oppressed and rack-rented. Land-jobbers and speculators obtained leases for the purpose of re-selling them at a profit. In order to recoup their outlay on reclamation, the original lessees often sublet to smaller lessees in return for cash payments, and the same process was carried down lower in the chain, with the result that the land was actually reclaimed and cultivated by peasant cultivators paying rack rents. Some of the lessees were in the habit of inducing tenants to take up land for reclamation on easy terms under invalid documents, and of ousting, or attempting to oust them, in favour of a new set of tenants as soon as the land had been brought under cultivation. Others neglected to repair the embankments which they were bound to maintain under the leases granted to them, and which alone protect the lands of their tenants from the ingress of salt water. Others again enhanced the rents by consolidating *ābwābs* with them and refused to give *pattās*, or rent receipts, except under illegal and oppressive conditions.

In these circumstances, it was decided in 1904 to abandon the system of grants to large capitalists and to introduce *raiyatwāri* settlements as an experimental measure, small areas (maximum 75 *bighas* and minimum 10 *bighas*) being let out to actual cultivators, whom Government assisted by means of advances, by constructing tanks and embankments, and by clearing jungle. The experiment proved a success in Bakargunge, where the cultivators as a rule clear the jungle themselves, but was a failure in the 24-Parganas, where the areas available for settlement present greater difficulties in the way of reclamation. Direct reclamation and *raiyatwāri* settlement were tried on a large scale in Fraserganj, but the reclamation proved unexpectedly expensive, and cultivators could not be induced to settle on the island on remunerative terms. In other resumed estates the results of direct reclamation and *raiyatwāri* settlement were equally discouraging, and in 1910 it was decided that the experiment must be abandoned and the old system of leases to large capitalists reverted to A new set of rules for settlement with large capitalists is under consideration, the object in view being to avoid, as far as possible, a recurrence of the abuses which grew up under the old capitalists rules.

Sāgar Island. A separate set of rules was issued in 1897 for the grant of waste lands in Sāgar Island. The waste lands to which these rules apply include all the ungranted and unoccupied land not contained within the boundaries of six grants, viz., Mud Point, Ferintosh, Bāmankhāli, Trower Land, Shikārpur and Dhobelāt or Gangā-Sāgar, which were assigned rent-free in perpetuity on condition that the grantees constructed protective works. The Sāgar Island rules generally followed the rules for large capitalists in the Sundarbans, except that a rent-free period of 15 years was allowed, and that failure to clear one-eighth of the entire grant by the end of the fifth year rendered the lessee liable, at the discretion of Government, to forfeiture of the lease or to a penalty of 4 annas per acre on the whole area for each year in which the area rendered fit for cultivation fell short of the area required to be cleared. The assessment was fixed at the rate of 2 annas per *bigha* from the 16th to the 20th year, and at the rate of 4 annas per *bigha* from the 21st to the 40th year. After the expiration of 40 years, re-settlements were to be made for periods of 20 years. The maximum area of a grant was fixed at 10,000 *bighas*. Leases were sold at an upset price of 8 annas an acre when there was one applicant, and to the highest bidder when there were more than one, the lessees being bound to construct protective works and to keep them in repair. Six leases have been granted covering an area of 20,362 acres; the rent payable is Rs. 956, which will eventually rise to Rs. 11,391.

These rules were suspended in 1904, when it was decided to introduce a system of *raiyatwāri* settlement. The *raiyatwāri* settlement rules have now again been suspended, and a modified set of rules for large capitalists is under consideration.

LAND TENURES. The proprietors of estates have freely exercised the power of alienation and have created a large number of tenures such as *patnīs*, *ijārās* and *gānthīs*. In creating these tenures, and even in giving a lease for a term of years, it has been and is a common practice for the tenure-holder to pay a bonus or premium. The system, while meeting the zamīndār's present necessity, means a loss to his posterity, because it is clear that if the bonus were not exacted, a higher rental could be obtained permanently from the land. The process of sub-infeudation has not terminated with the *patnīdārs*, *ijārādārs* and *gānthīdārs*. There are lower gradations of tenures under them called *darpatnīs*, *dārijārās* and *dargānthīs*, and even further subordinate tenures called *sepatnīs*, *segānthīs*, etc.

The *patnī tāluk*, or tenure, and the *ijārā*, or ordinary lease to a middleman, are too well known to require explanation. The word *gānthī* means assigned or allotted, and probably such tenures were originally created by the zamīndārs for the reclamation of waste land; but in process of time the term came to be applied to any tenure held immediately under a proprietor or *tālukdār*. It is now chiefly used to denote tenures or under-tenures held by persons who sublet to cultivators the whole or the greater portion of their lands. *Gānthi.*

Chakdāri is a term applied to leases, generally to middlemen who pay a fixed rent to the superior landlord and make their profit by subletting the land to cultivators. A *chak* is a plot of land situated between well-defined boundaries. It may be of any size, and sometimes in the Sundarbans consists of thousands of *bighas* of land. In the Sundarbans, and estates bordering on them, these *chaks* are often leased out for reclamation. The conditions are similar to those imposed by Government. The lease granted is of a permanent character, with a rent-free period and then a rent fixed at progressive rates. The tenure is liable to forfeiture for breach of the clearance conditions, and is held immediately under the zamīndār or a grantee of Government. It is a saleable and transferable tenure, and the *chakdār* undertakes to construct the necessary embankments for the protection of his *chak*, and to keep them in repair at his own expense when constructed. *Chakdār-leases.*

Non-agricultural tenures granted for building purposes to traders, artisans and other non-agricultural classes are called *khānābāri* and are numerous in the district. *Khānā-bāri.*

There are a large number of permanent tenures in the Government estates of Barnagore and Panchānnagrām. The holders pay rent to Government at fixed rates, and their tenures are saleable and heritable. They are of very old standing, and it is not known how they originated, but it is probable that they sprung up through the sufferance of Government. Permanent tenures of Barnagore and Panchānnagrām.

The revenue history of the Sundarbans is distinct from that of the rest of the district, and presents several peculiar features, so that a separate account of it is necessary.* REVENUE HISTORY OF THE SUNDARBANS.

In the first place, it must be explained that when the East India Company acquired the *Dīwāni* or civil administration of Early schemes of reclamation.

* This summary has been reproduced almost entirely from an article by Mr. F. E. Pargiter, I.C.S., *Cameos of Indian Districts.—The Sundarbans*, published in the Calcutta Review, October 1889. A full account will be found in *The Revenue History of the Sundarbans from 1765 to 1870* (Calcutta 1885) by the same author.

Bengal in 1765, the Sundarbans extended much further north than at present. Even in the vicinity of Calcutta the country was largely an uncultivated waste, especially to the east, where the forest approached to within about seven miles of the town. The Sundarbans, in fact, constituted a "no man's land;" it was covered with uncleared forest, and the forest was admittedly the property of the State.

The first effort to reclaim it was made in 1770 by Mr. Claude Russell, the Collector-General, in the present district of the 24-Parganas. He granted leases which allowed the lessees an initial period free of rent until they should have made some progress in cultivation, and fixed an ultimate rate of about Re 1-8 per acre on all the lands that might be found reclaimed when surveys should be made. These lands were called *patitābādi tālu s*, *i e.*, tenures for the reclamation of waste land. Very considerable progress was made by the lessees, and the neighbouring zamīndārs also busied themselves in promoting cultivation, so that during the next forty years the country was cleared almost down to Sāgar Island on the south, and nearly as far as Port Canning on the east. The next effort was made by Mr. Tilman Henckell, Judge and Magistrate of Jessore, who proposed to lease out small plots to raiyats (on terms very similar to those of Mr. Claude Russell's leases) so as to establish a body of independent peasant proprietors holding directly under Government. The scheme was approved by Warren Hastings, and Mr. Henckell, after roughly defining the boundaries of the forest, granted about 150 leases during 1785.

The scheme, was however, not a success, for it was opposed by all the neighbouring zamīndārs, who claimed the lands cleared by the grantees, and indeed all the forests as far as the sea, but declined to give Henckell any information about their estates that might enable him to decide the disputes. During 1786 he marked off by bamboo stakes the line which he took to be the northern limit of the Sundarbans and the southern boundary of the zamīndāri lands; and this strengthened the position of the lessees, but in the end the zamīndārs proved too strong for the new settlers. In 1792 they had all disappeared except sixteen, and in their case the character of the scheme had been modified, for the lessees developed into *tālukdārs*, their lands being called Henckell's *tāluks*.

Surveys.

About the year 1810 various schemes appear to have been broached for the improvement of the Port of Calcutta. One was to reclaim Sāgar Island; another was to construct wet docks at Diamond Harbour; and in 1816 it was even proposed to construct a canal, 47 miles long, from Calcutta to Channel Creek, the cost of which was so under-estimated that it was put down at only 10 lakhs. These schemes drew attention to the Sundarbans, and various surveys were carried out.

Captain Robertson surveyed the main water-routes from the Hooghly as far as the district of Noākhāli during 1810; and a portion of the sea coast, east of the Hooghly, was surveyed by Lieutenant Blane in 1813-14. The Sundarbans (exclusive of the sea face) from the Hooghly as far as the river Pasar, were surveyed by Lieutenant W. E. Morrieson during the years 1811-14, and his results were corrected by his brother Captain Hugh Morrieson in 1818. This was a great work, carried out in spite of many difficulties and dangers, and has been the basis of all subsequent maps of the Sundarbans.

Legislation.

In 1814 the Court of Directors directed that settlements should be concluded with the actual occupiers for the lands already brought under cultivation, while holding out reasonable encouragements for further reclamation. An attempt was accordingly made during the years 1814-16 to re-measure the grants already made and to revise their rentals, but it met with only partial success. The advantages, however, that the State might gain from the opening up of the Sundarbans were clearly perceived, and a law was passed in 1816 (Regulation IX of 1816), sanctioning the appointment of an officer to deal with the Sundarbans, to be styled the "Commissioner in the Sundarbans," with all the powers and duties of a Collector.

Mr. D. Scott was the first Commissioner, and he began inquiries and measurements in the country south and east of Calcutta. It was then found that encroachment and reclamation had been steadily and continuously carried on, partly by the lessees, partly by the zamīndārs, and partly by other unauthorized persons. All the new land brought under cultivation was held without payment of any revenue to the State. The proposal to levy revenue from it naturally aroused the opposition of all the persons interested, and especially of the zamīndārs, who claimed the whole of the forest. They resisted the operations with force, fraud and chicanery, so much so that Mr. Scott had to be allowed an esc rt of twelve sepoys for his personal protection. In 1817 Government expressly declared in a law then enacted (Regulation XXIII) that the Sundarbans were the property of the State, and asserted its right to the revenue of lands not included within the boundaries of estates for which a settlement had been made. In spite of this, it began to entertain misgivings as to whether it was not debarred by the Permanent Settlement from dealing with such lands. This hesitation, the intricacy of the questions that arose, and the difficulties experienced practically put an end to the operations for the next four years.

In 1821 the Sundarbans office was reconstituted under Mr. Dale, and was reinforced by a survey party under Ensign Prinsep, with the wider object of demarcating the State lands from private estates. But directly the inquiries began in the district of the 24-Parganas, it was found that, if the zamīndārs were to be believed, there were no State lands at all, for they claimed all the forest that abutted on their estates down to the sea coast, and yet declined to point out their lands. The only course, therefore, was to survey all the lands that had been brought into cultivation during the previous thirty years, and that was done. Mr. Prinsep surveyed the line of forest from the river Jamuna to the Hooghly in 1822 and 1823; and, with the aid of the Morriesons' map, divided all the forest lands between those rivers into blocks and numbered them. This was the beginning of the Sundarbans "lots."

Attention was next directed to the claim of the State to demand revenue both from the recently reclaimed lands and from the forest. It was beyond doubt that the land had been reclaimed since the Permanent Settlement, but those in possession claimed to hold both land and forest as part and parcel of their estates, at the revenue fixed at the Permanent Settlement, and free from increased revenue. Resumption (*i.e.*, the establishment of the right of the State to demand revenue from lands that pay no revenue) was not an easy matter, on account of the intricacy of the claims, the paucity of trustworthy documents and the fabrication of false papers; but the exertions of Messrs Dale, R. D. Mangles, John Lowis and other officers overcame the difficulties, and by 1828 the State had recovered all the lands that had been surreptitiously encroached on, and all the forest in the 24-Parganas. In that year a final declaration of the rights of the State over the recent cultivation and the forest was made in Regulation III, in which it was laid down—"The uninhabited tract known by the name of the Sundarbans has ever been, and is hereby declared still to be, the property of the State; the same not having been alienated or assigned to zamīndārs, or included in any way in the arrangements of the Perpetual Settlement. It shall therefore be competent to the Governor-General in Council to make, as heretofore, grants, assignments and leases of any part of the said Sundarbans, and to take such measures for the clearance and cultivation of the tract as he may deem proper and expedient." It also enacted that the boundary of the Sundarbans forest should be determined by the Sundarbans Commissioner and laid down by accurate survey.

Mr. William Dampier was now appointed Commissioner, and Lieutenant Hodges Surveyor, their jurisdiction being extended over the whole of the Sundarbans in Khulna and Backergunge. They defined and surveyed the line of forest from the Jamuna (where one end of Prinsep's line was) up to the eastern limit of the Sundarbans, during the years 1829 and 1830; and Mr. Dampier formally affirmed Prinsep's line in the 24-Parganas in 1832-33. "Prinsep's line" and "Hodges' line" are the authoritative limits of the then Sundarbans forest, while the map prepared by Lieutenant Hodges in 1831, from his own surveys and those made by his predecessors, has been the standard map of the Sundarbans ever since. Following Prinsep's plan, he divided all the forests as far as the river Passur into blocks, and, revising the numbering, reduced the whole of his and Prinsep's blocks into a series numbered from 1 to 236. The aggregate area of these 236 Sundarbans "lots" was computed at 1,702,420 acres, or 2,660 square miles. Beyond the river Passur no detailed survey of the forest had been made, and it was impossible to continue the allotments there. Sundar-bans lots.

When the forest line had been authoritatively determined, it became necessary to deal with the lands already reclaimed which were held free of land revenue, *i.e.*, to resume and settle them. The resumptions may be divided into two stages. The proceedings till 1836 dealt with patent and incontestable cases of encroachment, and very large tracts were recovered to Government. After that year the proceedings were more intricate and difficult, for inquiries were then systematically conducted into the smaller and clandestine cases of encroachment, which had either escaped notice before, or had been left over, because of their difficulty, for more thorough and leisurely investigation. As fast as each estate was resumed, it was brought under settlement, and the increase of revenue may be computed roughly at about 2½ or 3 lakhs up to about the year 1844. Resumptions.

As early as the year 1819 the Government had contemplated making grants of the forest lands with a view to their being cleared. Nothing definite, however, was decided upon, though a few isolated grants were made both before and after that year, and it was not till 1830 that rules for the grant of the forest were issued. Applications poured in mostly from Europeans resident in Calcutta, who formed sanguine expectations of success. With the exception of some lands reserved for the Salt Department, applicants practically got gratis whatever they Rules of 1830.

asked for in the 24-Parganas and Khulna. During the two years 1830 and 1831, 98 lots were granted away, and 12 more during the more next five years, with a total area of 551,520 acres. These grants were made in perpetuity at a rental of about Re. 1-8 per acre, and nothing was payable during the first 20 years; but it was stipulated that one-fourth of the area should be rendered fit for cultivation within five years, under pain of the grant being forfeited to Government, this condition being held necessary to ensure that the grantees should carry out the object for which they have received the lands After the first eager competition, the applications fell off, but they revived in 1839, and about half of the forfeited grants were leased out again besides some twelve new lots.

Though such a large area was leased out, a considerable portion was never brought under cultivation, and consequently the grants lapsed. Some of the persons who got grants were mere speculators, who did not attempt to clear their lands, but realized whatever profit they could get from the wood and other natural products, and sold the lots as soon as they could find a purchaser. In a few cases the grantees complied at once with the condition requiring clearance, but in rather more than one-third of the lots, the difficulties and losses of the grantees proved insuperable, so that the Government was obliged to cancel their grants. Few grantees were able to succeed who had not ample capital at their command. The work of reclamation required unceasing care and vigilance; it was liable to be interrupted at any moment by the desertion of the raiyats, and fresh men had to be engaged at heavy expense. If an embankment chanced to be breached, salt water poured in and ruined the soil with a deposit of salt. Moreover, as each lot was given to the first applicant, the best lots were taken up by the earliest applicants, and until some progress was made in the lots bordering on the cultivated tracts it was almost hopeless for a grantee whose land lay deeper in the forest to succeed in his undertaking.

Rules of 1853. In these circumstances the grantees begged for more liberal terms, and Government, which had already had to relax the terms in individual cases by waiving the right of forfeiture, recognized the justice of their demand. New rules were accordingly published in 1853, of which the chief conditions were:—(1) Grants were to be made for 99 years, and were sold to the highest bidder if there was competition. (2) The revenue assessed on them was reduced to about 6 annas per acre and did not become payable till the 51st year, after a long and very gradual enhancement commencing from the 21st year. (3) Reclamation was more carefully provided for, the grantee being required to have one-eighth of his grant fit for cultivation in five years, one-fourth in ten, one-half in twenty, and practically he whole in 30 years, under pain of forfeiture. The earlier grantees were allowed the option of giving up their old leases and taking fresh leases under the new rules. This boon was greatly appreciated, and about seventy of the earlier grantees accepted it and commuted their leases. The new rules gave a fresh stimulus to the work of reclamation, but in spite of the greater liberality of the conditions regulating clearance, they were not complied with, and during the following ten years about seventy grants fell in through forfeiture; from these and other lands about ninety fresh grants were made. Still, it was far from the policy or practice of Government to press the conditions harshly; and if a grantee exerted himself and yet failed through causes beyond his control, either forfeiture was waived, or he was allowed further time to satisfy the conditions.

Waste land rules

After the Mutiny, two proposals were brought forward for the disposal of waste lands generally. One was to sell them outright, exempt from land revenue, the other was to allow land-owners to redeem their existing land revenue by paying it off once for all by one capitalized sum. These measures were generally advocated with the object of promoting the settlement of Europeans in India; and after much discussion, rules, called the Waste Land Rules, which embodied these views, were issued in 1863. Some thirteen lots were sold under these rules in 1865 and 1866, but many of the purchasers were indifferent, or were unable to complete the purchase money during the ten years allowed for the payment by instalments, and eight of those lots came back to Government in subsequent years. The rules for the redemption of land revenue, however, met with more success, and were taken advantage of in some twenty lots.

Schiller's scheme.

The granting out of lands had been stopped meanwhile, but the new measures were found fault with on the ground that the upset price was excessive, and that only capitalists could take advantage of them. Applicants for land, therefore, preferred the grant rules of 1853, and wished to go back to them, but no definite change was made, for Mr. F. Schiller's great scheme of Sundarbans reclamation was then before the public. He and eight other gentlemen, European and Indian, applied to Government in 1865 for the purchase of all the remaining ungranted waste lands, proposing to raise a capital of not less than one million sterling, and to reclaim the lands by means of labour imported from China, Madras, Zanzibar and other places. The Government was prepared to sanction this ambitious scheme, but declined to countenance any proposal to import labour from Africa. Mr. Schiller then attempted to start a company with a capital of nearly two millions, and continued his efforts for three years, but so much money had been lost in Sundarbans enterprises, that people had grown cautious and the public could not be induced to join. His efforts proved unsuccessful, and the scheme at length fell through in 1868, when the Port Canning scheme was competing for the support of the public.

Port Canning scheme.

About the year 1853 the idea was started of making a subsidiary port to Calcutta on the river Mātla. The river was surveyed, and Government bought up for Rs. 11,000 lot 54 (at the head of the river on the west side) with an area of about 8,260 acres, for the purpose of constructing a ship canal and railway to connect the river with the Hooghly. The lot was only partially cleared along the river frontage, and this portion was surveyed for six miles, and marked out into roads and "lots" for the construction of the new town and port. Measures were also taken to clear the remainder of the lot and people it with raiyats. This was a tedious and expensive undertaking, and seems to have occupied about seven years. The establishment of the port was begun about 1858, and it was called Port Canning. The lands on the river bank were the town lots, and all the rest remained agricultural lands. Leasehold rights in the town lots were sold at public auction for a term of sixty years, and were largely bought up. In 1862 the Port Canning Municipality was formed and formally obtained from the Government its property in the town lands. Attempts were made to raise public loans for the improvement of the town and port, but they were not attended with success. In connection with this scheme a company was started, called the "Port Canning Land Investment, Reclamation and Dock Company, Limited," for the purpose of purchasing and reclaiming the waste lands on the river Mātla. The company bought seven lots, and held ten others in grant, and it contributed largely to a debenture loan that the municipality succeeded in raising in 1865. But that body needed more money next year, and Government lent it half a lakh without interest for five years, retaining the port dues in its own hands. A railway was constructed between Calcutta and Port Canning, and wharves were built in connection with the railway; but the port failed to attract trade, and the scheme failed. The municipality got into hopeless pecuniary difficulties, which at length brought it into costly litigation with the Port Canning Company and with Government. The final result was that many of the lease-holders of the town lots, who held large quantities of the municipality's debentures, commuted their debentures for the free-hold right in their lots; and Government bought up the whole of the property of the municipality in the Civil Court, and paid off all the other debenture-holders.

As already mentioned, it was proposed about 1810 to clear Sāgar Island, in order to benefit the navigation of the Hooghly. Two persons tried in turn but failed, and many others applied for grants in the island, but it appeared they had no real intention of clearing the land, and only wanted to secure a nominal property that might eventually become valuable. The island was surveyed by Lieutenant Blane in 1813-14 and its area was computed at about 143,550 acres. The Collector of the 24-Parganas, Mr. Trower, began clearing in a central portion, which was named, after him, Trowerland; but he found that cultivation could be undertaken better by private persons, and convened a meeting of merchants and others in Calcutta in 1818. The result was that a company was started, called the "Sāgar Island Society," with a capital of 2½ lakhs. The Government granted the island to the company in perpetuity, the first 30 years being free of rent, with various stipulations as to clearing, etc., failure in which would entail forfeiture. The management was in the hand of 13 trustees—a number which to the superstitious will seem to have prophesied misfortune from the beginning. The society began energetically, and very satisfactory progress was made in the four northern portions called Mud Point, Ferintosh, Trowerland and Shikārpur, and in a portion at the extreme south called Dhobelāt, until May 1833, when a great gale and inundation occurred, which destroyed almost every thing and compelled the society to throw up the scheme in despair. Four gentlemen, Messrs. Hare, Macpherson, Hunter and Campbell, then bought the four northern portions and carried on the undertaking. Government also conferred on them the privilege of making salt, from which, and from rice cultivation combined, they reaped a lucrative return. The island continued in fair prosperity, although storms occurred in June 1842, October 1848 and June 1852; and the rent-free term was extended to 1863. In the latter year the northern portions had been more or less cleared, and there was also some cultivation found in Dhobelāt and another portion; all the rest of the island was jungle, except the spot occupied by the light-house. Next year all the work was undone by the cyclone and storm-wave of 5th October, 1864;

Sāgar Island.

and another storm on the 1st and 2nd November, 1867, threw the island back still further. These calamities forced attention to the necessity of providing means of safety, especially for human life, for it was patent that the embankments already erected had not been a sufficient protection. After much discussion it was settled that, in each estate in the island, a central place of refuge should be constructed, consisting of a tank surrounded by an embankment 16½ feet high, that no habitation should ordinarily be built more than a mile from a place of refuge, and that embanked paths should be made connecting the places of refuge with the houses. Subject to these and minor conditions, the cultivated lands in the five estates already mentioned were granted free of rent in perpetuity in 1875.

CHAPTER XII.

GENERAL ADMINISTRATION.

SUB-DIVISION AND STAFF.

For administrative purposes the district is divided into five subdivisions with headquarters at Alipore, Barrackpore, Bārāset, Basīrhāt and Diamond Harbour, the area and population of which, according to the census of 1911, are shown in the margin. The area known as the Suburbs of Calcutta, *i.e.*, the municipalities of Cossipur-Chitpur, Māniktala and Garden Reach, is geographically part of the Alipore or Sadar subdivision, but is classed separately, as it is treated as a distinct unit for census purposes, and its administration is, in several respects, different from that of the rest of the district. Police jurisdiction in the three towns is exercised by the Commissioner of Police, Calcutta, and the police force is drawn from the Calcutta Police, while cases occurring within their limits are tried by Deputy Magistrates, subordinate to the District Magistrate of the 24-Parganas, who hold their Courts at Alipore and Sealdah.

Subdivision.	Area in square miles.	Population.
Alipore ...	1,164	753,348
Barrackpore ...	190	292,524
Bārāset ...	275	292,791
Basīrhāt ...	1,922	429,476
Diamond Harbour	1,283	515,725
Suburbs of Calcutta.	10	147,240

For some administrative purposes the "Added Area" of Calcutta forms part of the district. This, it may be explained, is the portion of Calcutta lying east and south of the "Old Town" (between the Circular Road and the Hooghly) and separated from it by the Circular Road. It includes seven entire wards, viz., Entally, Beniāpukur, Ballygunge-Tollygunge, Bhawānipur, Alipore, Ikbālpur and Watganj (in which Kidderpore is situated), and also parts of four other wards, which are known collectively as the "Fringe Area." The latter is a strip of land lying between the Upper Circular Road and the Circular Canal, which forms part of Ward No. 1 (Shāmpukur), Ward No. 3 (Bartala), Ward No. 4 (Sukea Street), and Ward No. 9

(Muchipāra); it is known as Ultadānga in Ward No. 1, as Māniktala in Ward No. 3 and as Baliāghāta in Wards Nos. 4 and 9.

The "Added Area" forms part of the 24-Parganas for civil and criminal purposes, criminal cases arising within its limits being tried by the Police Courts at Sealdah and Alipore, and committed, if necessary, to the Sessions Court at Alipore; but for municipal and police purposes it is an integral part of Calcutta, for its municipal administration is entirely under the Corporation of Calcutta and police jurisdiction is exercised by the Commissioner of Police, Calcutta. It also belongs to the 24-Parganas for all revenue purposes except excise, *i.e.*, its revenue administration, except as regards excise, is under the District Officer. The tract east of Tolly's Nullah and the Lower Circular Road, which comprises Bhawānipur, Ballygunge and Entally, is included in the Panchānnagrām Government Estate, and to the east of Tolly's Nullah a considerable area is included in the Sāhibān Bāgicha Government Estate, the remainder being permanently settled estates belonging to private owners.

At the headquarters station the District Magistrate is assisted by a large staff of Deputy Magistrates, Probationary Deputy Collectors, Sub-Deputy Collectors and Probationary Sub-Deputy Collectors, while each of the Subdivisional Officers has a Sub-Deputy Collector under him, the Subdivisional Officer of Diamond Harbour also having the services of a Deputy Collector. The sanctioned staff of Deputy Magistrates at Alipore consists of eight Deputy Magistrates of the first class and one Deputy Magistrate of the second or third class.

The Collector is *ex-officio* Collector of Calcutta, but revenue (which in the "Old Town" is really ground rent* permanently fixed at Rs. 18,163) is collected by an officer, styled the Deputy Collector of Land Revenue, Calcutta, who is also Collector of Stamp Revenue in the "Old Town" and Superintendent of Excise Revenue in Calcutta, and in so much of the district as is under the jurisdiction of the Commissioner of Police, Calcutta, *i.e.*, in the municipalities of Cossipur-Chitpur, Mānicktala and Garden Reach: in his functions as Deputy Collector of Land Revenue this officers is subordinate to the Collector of the 24-Parganas. In his capacity as District Magistrate, the Collector is *ex-officio* visitor of the lunatic asylums at Bhawānipur and Dullānda, and is a member of the Board constituted for the Prince Albert Victor

* In the "Old Town" there is, strictly speaking, no land revenue, as in 1758 the East India Company obtained from the Nawāb of Bengal a revenue-free grant of the area on which the city now stands.

Asylum for Lepers at Gobra. He is also vested with the powers of an Assistant Commissioner under the Indian Salt Act: there is an Assistant Commissioner of the salt preventive force for this district and Khulna. The Subdivisional Officer of Barrackpore is *ex-officio* Inspector of Factories within his subdivision and the senior covenanted assistant or Joint Magistrate at Alipore outside it.

Civil justice is administered by the District and Sessions Judge, two Additional District and Sessions Judges (of whom one is also Additional District and Sessions Judge of Hooghly, and the other is a special Land Acquisition Judge), three Sub-Judges, a Judge of the Small Cause Court at Sealdah and 16 Munsifs. Three of the Munsifs are stationed at Alipore, three at Bāruipur (one being an Additional Munsif), four at Diamond Harbour, and two each at Bārāset, Basīrhāt and Sealdah. Both the District Judge and the Additional Judge have jurisdiction in Calcutta under the Land Acquisition Act. The civil work of the district is heavy, as may be judged from the fact that in 1911 there were 29,268 suits disposed of under the ordinary procedure, and 14,598 under the Small Cause Court procedure, while 299 appeals were disposed of by Sub-Judges and 410 by District Judges. CIVIL JUSTICE.

In addition to the Stipendiary Magistrates, there is an Honorary Magistrate at Ichāpur and 25 benches of Honorary Magistrates at the places shown below:— CRIMINAL JUSTICE.

Place.	Number of Magistrates on Bench.	Place.	Number of Magistrates on Bench.
Alipore ...	... 21	Gobardānga	... 5
Bāduria ..	... 6	Hālishahar	... 6
Bārāsat	... 8	Jaynagar ...	... 6
Barānagar	... 10	Kānchrāpāra	... 5
Barrackpore	... 10	Khardah ...	... 8
Bāruipur ...	... 6	Naihāti ..	... 11
Basīrhāt ...	... 5	Palta ...	... 5
Bhātpāra ...	... 9	Pānihāti ...	... 5
Budge-Budge	... 7	Rājpur ...	... 7
Cossipur-Chitpur	... 12	Sealdah ...	... 22
Diamond Harbour	... 4	Tāki ...	... 1
Dum-Dum	... 7	Titāgarh ...	... 11
Gārulia ...	... 5		

There are also Police Court Magistrates at Alipore and Sealdah, and a Cantonment Magistrate for the cantonments at Barrackpore and Dum-Dum. The Sessions Judge and the District

Magistrate have criminal jurisdiction in the "Added Area" of Calcutta. The Deputy Magistrate of Sealdah tries cases from the municipalities of Cossipur-Chitpur and Māniktala, and from the Entally, Beniāpukur, Baliaghāta and Ultadānga thanas of Calcutta. The Deputy Magistrate of Alipore tries cases from Garden Reach and the other thanas of the "Added Area," viz., Alipore, Ballygunge, Tollygunge, Bhawānipur, Ikbālpur and Watganj.

The number of criminal cases disposed of in 1911 was 28,466, or double the number disposed of ten years previously.

POLICE. There are altogether 41 thanas in the district, of which 27, as shown below, are under the District Superintendent of Police:—

Thana.	Police-station.	Thana.	Police-station.
Sadar Subdivision.		*Barāse Subdivision.*	
Bāruipur ...	Bāruipur.	Bārāsat ...	Amdānga.
	Pratābnagar.		Bārāset.
Behāla ...	Behāla.		Rājārhāt.
	Maheshtala.	Deganga ...	Deganga.
	Mātiābruz.	Hābra ...	Hābra.
Bhāngar ...	Bhāngar.	*Barrackpore Subdivision.*	
Budge-Budge ...	Budge-Budge.	Barānagar ...	Barānagar.
Jaynagar ...	Jaynagar.	Barrackpore ...	Barrackpore.
Mātla (Canning) ...	Mātla (Canning).		Titāgarh.
Sonārpur ...	Sonārpur.	Dum-Dum ...	Dum-Dum.
Tollygunge ...	Tollygunge.	Khardah ...	Khardah.
Vishnupur ...	Vishnupur.	Naihāti ...	Jagatdal.
			Naihāti.
		Noāpāra ...	Noāpāra.
Diamond Harbour Subdivision.		*Basirhāt Subdivision.*	
Diamond Harbour	Diamond Harbour.	Bāduria ...	Bāduria.
Falta	Falta.		Sarupnagar.
Kulpi ...	Kulpi.	Basirhāt ...	Basirhāt.
	Kākdwīp.	Hāroa ...	Hāroa.
	Sāgar.	Hāsanābād ...	Chaital.
Magrā Hāt ...	Magrā Hāt		Hāsanābād.
Mathurāpur ...	Mathurāpur.		Sandeshkhāli.

N.B.—The thanas of Magrā Hāt, Falta and Budge-Budge were formerly known as Bānkipur, Debīpur and Achipur respectively.

The police in the Suburbs of Calcutta are under the orders of the Commissioner of Police, Calcutta, whose jurisdiction embraces the Cossipur-Chitpur and Māniktala municipalities and part of the Garden Reach municipality, as well as the Added Area" east and south of the Circular Road and Tolly's Nullah, which lies within the district limits, but forms part of the Calcutta municipality. The Suburbs are divided into two police divisions, the northern and southern, each under a Police Superintendent, and include the 14 thanas shown in the margin. According to the returns for 1911, the sanctioned strength of the police under the District Superintendent is: one Assistant Superintendent, nine Inspectors, 98 Sub-Inspectors, three Sergeants, 161 head constables and 1,277 constables. The rural force in the same year consisted of 3,072 chaukīdārs and 288 dafadārs.

Alipore.	Entally.
Baliāghāta.	Garden Reach.
Ballygunge.	Ikbālpur.
Beniāpukur.	Mānicktollah.
Bhawānipur.	Tollygunge.
Chitpur.	Ultadānga.
Cossipur.	Watganj.

JAILS.

There are seven jails in the district, as shown in the marginal statement, which also shows the number of prisoners for which there was accommodation in the year 1911.

Place.	Class or name.	Accommodation.
Alipore ...	District and Central	1,850
Do. ...	Juvenile ...	270
Kālīghāt ...	New Central ...	1,481
Bārāset ...	District ...	174
Barrackpore ...	Subsidiary ...	42
Basīrhāt ...	Ditto ...	12
Diamond Harbour	Ditto ...	37

The jail first mentioned, which is said to have been built in the time of Lord Cornwallis, is generally known as the Alipore Central Jail. It is, however, a district as well as a Central Jail, and it receives local convicts and under-trial prisoners from the Courts at Alipore, as well as convicts sentenced to long terms of imprisonment; to be more precise, prisoners sentenced to imprisonment for more than two years are transferred to it from other jails, most of them in the Presidency Division. Prisoners sentenced to transportation to the Andamans are also sent there, prior to transhipment, from Bengal, Bihār and Orissa, the United Provinces, the Central Provinces, the Punjab, Assam and even Bombay. Batches of them are inspected every six weeks by a standing transportation committee before being embarked for Port Blair. The jail has two chief industries, viz., the jute mill and the iron-works. The jute mill supplies gunny bags to the Transport and Supply Department of the army, and to other jails, the bags supplied to the latter being used for the storage of the grain provided for prisoners' diet. The iron-works manufacture a large variety of articles, such as Larymore boilers for boiling water (which are supplied to all jails and to a few cantonments), water cans, water reservoirs, hospital iron beds (for jail and municipal hospitals, jail warders and the police), water carts, cooking ranges of the special jail pattern, cooking pots and *degchis*, night-soil trucks, whipping triangles, milk cans, wire-netting, pumps for wells, incinerators, office furniture, almirahs and chairs, oil mills of the jail pattern, etc. These are supplied to other jails and to a few Government departments and public bodies.

The New Central Jail is situated on the banks of Tolly's Nullah close to the iron Kālighāt bridge at Alipore. It was built to replace the old Presidency Jail on the Calcutta Maidān, the site of which was required for the erection of the Victoria Memorial Hall. It is a modern up-to-date prison with modern improvements, including electric lighting and a water-closet system. It is admirably designed and has only one defect, *viz*, the limited space available for its numerous buildings. Its industries were taken over from the old Presidency Jail. The chief industry is printing, some 800 convicts being employed daily in printing and binding the forms and registers required for Government offices in Bengal and Assam. Close by the jail there is a large block called the Forms Block, which is to be amalgamated with the Jail Press. The New Central Jail will then not only print the forms, but also receive the indents for them and distribute forms, registers, etc., to all offices in the Bengal Presidency and Assam. This will not only lead to a more prompt supply of forms, but, as convict labour will, to some extent, replace the paid labour now employed by the separate Forms Department, it is expected to result in the saving of a considerable sum to Government.

As the new jail is situated outside the limits of the Presidency town of Calcutta, the question of the jurisdiction of the High Court has arisen, and it has been decided that its jurisdiction should be extended so as to include the road leading to, and the area occupied by, the old Central Jail at Alipore. Accordingly, by a notification issued in October, 1913, the old Alipore Central Jail has become the Presidency Jail of Calcutta (*i.e.*, it will receive prisoners direct from the High Court and the Presidency Courts of the town of Calcutta), while the name of the New Central Jail at Kalighat has been changed to the Alipore Central Jail. Quarters for European under-trial prisoners and for civil debtors have recently been built in the old Central Jail to enable it to fulfil its functions as the Presidency Jail of Calcutta.

The Juvenile Jail at Alipore is the pioneer Juvenile Jail in India, and was opened in November 1908, its object being to put into practice a scheme for the reform of juvenile prisoners suggested by Lieutenant-Colonel W. J. Buchanan, C.I.E., I.M.S., as the result of experience gained by visits to the celebrated Borstal Jail for "juvenile adult" offenders near Rochester in England. It is accommodated in the buildings formerly used by the Alipore Reformatory School, which was removed to, and amalgamated with the Reformatory School, at Hazāribāgh. Various useful trades are taught, such as weaving, carpentry, tinsmith's work, etc.; in addition to this, the prisoners make up and despatch the preparations of cinchona and quinine supplied to Government institutions and to the public. The system of convict officers has been abolished, and paid warders or teachers are in charge of the gangs of juveniles. An endeavour is made to limit the admissions to youths with sentences of over six months, boys with shorter sentences being sent to the juvenile wards in each of the Central Jails in the Presidency. There is a system of rewards for good conduct, and selected boys are chosen as "star class boys" and "monitors." The system works well, but the great desideratum is a committee to find work for this class of youth on release: the difficulties in the way of arranging for the emigration of Indian boys are too obvious to require explanation. So far little has been done to meet this great need, so that the characteristic feature of the English Borstal system is wanting.

The following statement shows the charges of both branches of the Public Works Department in this district. The divisions are in charge of Executive Engineers, and the subdivisions are held by Subdivisional Officers, who may be either Assistant Engineers or Upper Subordinates :— PUBLIC WORKS DEPARTMENT.

Division.	Subdivision.
IRRIGATION.	
Northern Drainage and Embankment.	Surjipur. Diamond Harbour.
Circular and Eastern Canals ...	Calcutta Canals. Bhāngar.
ROADS AND BUILDINGS.	
First Calcutta	Maidān.
Second Calcutta	Belvedere and New Presidency Jail. Alipore. Dhullundah Barrackpore. Stores.

An area of 1,711 square miles in the Sundarbans is under the management of the Forest Department as "protected forest." The area under its control was formerly much larger and has diminished owing to the spread of cultivation. During the ten years ending in 1901-02, 421 square miles were leased out for cultivation, reducing the area of protected forest to 1,844 FORESTS.

square miles, and since then 133 square miles have been similarly deforested. The forests now maintained by the Forest Department are bounded on the south by the Bay of Bengal and extend from the eastern boundary of the district up to the cultivated lands west of the Thākurān river. Beyond that river only a few islands remain under forest, and doubtless these will eventually be brought under cultivation. The forest growth is greatly inferior to that found in the forests of the neighbouring district of Khulna, the reason generally assigned being that the rivers in the 24-Parganas have no longer a connection with fresh water streams, and are in reality arms of the sea, so that their water is saline throughout the year. *Sundrī* (*Heritiera littoralis*), which is so plentiful in the Khulna forests, here occurs as a stunted tree with extremely scanty distribution, except in a small area near the Raymangal estuary and close to the northern boundary. The principal species are *garān* (*Ceriops Candolleana*), *gengwa* or *geoa* (*Excoecaria Agallocha*) and *keora* (*Sonneratia apetatar*), of which the most important is the *garān* tree. The main sources of revenue are *garān* poles, *garān* fuel and other fuel. The minor products consist of the *hentāl* or wild date palm (*Phœnix palu losa*), *golpāta* or the leaves of the *Nipa fruticans*, which are used for thatching, *nal* and *hogla* reeds, which are made into matting, shells from which lime is produced, honey and bees-wax. The officer in charge of the forests is the Deputy Conservator of Forests in charge of the Sundarbans division, whose headquarters are at Khulna. The removal of produce is controlled by the ten marginally noted revenue stations commanding the principal rivers and *khāls*. At all these stations, except Basra, Mātla and Nārikeldānga, permits are issued to boats entering the Sundarbans to get wood or other forest produce.

Basanti.	Nalkara.
Basra.	Nārikeldānga.
Kākdwip.	Rāmpura.
Kultala.	Sāhebkhāli.
Mātla	Shikārpur.

REGISTRATION. There are 24 offices for the registration of deeds under Act III of 1887, as shown in the statement below, which also gives the salient statistics for the year 1912, when both registrations and receipts were the highest on record:—

Office.	Number of documents registered.	Receipts.	Expenditure.
		Rs.	Rs.
Alipore	3,807	22,803	12,726
Do. joint (Behāla) ...	4,027	5,120	2,369

Office.	Number of documents registered.	Receipts.	Expenditure.
		Rs.	Rs.
Bāduria	5,646	5,446	2,667
Barāset	5,526	4,784	2,045
Barrackpore	2,189	2,517	2,214
Bāruipur	4,584	4,668	2,983
Basīrhāt	5,530	6,677	2,666
Do. joint (Hasanābād) ...	2,658	2,860	2,142
Cossipur	1,041	2,510	2,161
Deganga	4,451	4,006	2,490
Diamond Harbour	5,437	5,172	2,437
Dum-Dum	1,910	1,862	1,738
Hābra ...	4,907	4,371	2,005
Jaynagar	6,558	6,235	2,730
Magrā Hāt (Hasura) ...	6,155	5,112	2,290
Ditto joint (Eyārpur) ...	2,718	2,175	1,860
Mātla	5,284	4,993	1,799
Mathurāpur	6,755	5,906	2,038
Naihāti	4,150	3,874	2,142
Sarār Hāt	3,498	2,791	1,631
Sealdah	4,358	8,935	3,198
Sultānpur (Lakhikāntapur) ...	3,710	3,221	2,701
Ditto joint (Tengra) ...	5,038	4,707	2,237
Vishnupur	4,910	4,421	2,205
Total ...	104,847	1,25,166	65,474

Details of the revenue of the district during the last decade will be found in the B Volume, which is published separately, and here it will be sufficient to say that the receipts in 1910-11 aggregated Rs. 49,29,952, representing an advance of 16 lakhs since 1901-02. The increase is chiefly due to the enhanced revenue from stamps, excise and "other sources"; the receipts from excise alone have been doubled in this period. The number of persons assessed to income-tax decreased from 4,055 to 1,813, owing to the amendment of the Act in 1903, by which the minimum income assessable to the tax was raised from Rs. 500 to Rs. 1,000. The number has since grown to 2,298, representing one in every 1,060 of the population. REVENUE.

Receipts in 1910-11.	Rs.
Land revenue ...	17,88,905
Stamps	11,09,936
Excise	10,83,741
Road and public works cesses	3,47,269
Income-tax	1,07,912
Other sources ...	4,92,189

CHAPTER XIII.

LOCAL SELF-GOVERNMENT.

DISTRICT BOARD.

The District Board consists of 21 members, of whom five are *ex-officio* members, six are appointed by Government and ten are elected. The District Magistrate is *ex-officio* Chairman. Details of the income and expenditure of the Board during the last ten years will be found in the B Volume of statistics. Both have been largely inflated during the last few years on account of the Magrā Hāt drainage scheme. As explained in a previous chapter, this is a scheme which provides for the improvement of the sanitary and agricultural condition of a tract of 290 square miles in the Diamond Harbour subdivision, at an estimated cost of 21 lakhs. Government has contributed five lakhs, and has undertaken the maintenance charges, in return for any income derivable from the scheme, including tolls to be levied on the drainage channels. The remainder of the cost is being met by the District Board from loans granted by Government, which will be repaid when recoveries are made from the proprietors of lands benefited by the scheme. In addition to this undertaking, the Board has guaranteed interest at the rate of 4 per cent., up to a maximum of Rs. 38,000, on the Bārāset-Basīrhāt Light Railway, and has taken a loan of Rs. 60,000 from Government for the construction of feeder roads to the line. Government has also contributed (in 1911-12) two lakhs towards a drainage project designed to improve a tract in the neighbourhood of Bārāset by opening up the Nawai and Sunthi rivers.

Excluding the sums received from Government as a loan for the Magrā Hāt scheme, but including grants from Provincial funds, the ordinary income of the Board is over 3 lakhs a year, as shown in the margin. The income in 1911-12 was Rs. 4,12,662, excluding the opening balance and including a sum of Rs. 1,40,000 received from Government for civil works. The Board maintains ten dispensaries and aids twelve others. It also maintains 3 middle schools and gives grants-in-aid to 48 middle, 66 upper primary, 1,032 lower primary and 110 other schools. The mileage of the roads kept up by it is:—metalled roads 179 miles, unmetalled roads 380 miles and village roads 1,042½ miles.

Year.		Income.
		Rs.
1906-07	...	3,18,528
1907-08	...	3,31,146
1908-09	...	3,12,295
1909-10	...	3,24,756
1910-11	...	4,16,072

LOCAL BOARDS.

There are five Local Boards, one having been constituted for each subdivision. The Sadar Local Board is composed of 16 members, that at Diamond Harbour has 12 members, at Bārāset 10 members, at Basīrhāt 9 members, and at Barrackpore 7 members. The Barrackpore and Bārāset Local Boards are presided over by the local Subdivisional Officers, and the other three Boards have non-official Chairmen elected by the members. These bodies have charge of unmetalled roads, village roads and pounds.

UNION COMMITTEES.

There are three unions in the district, viz., Bāsudebpur in the Bārāset subdivision and Itinda and Jadurhāti in the Basīrhāt subdivision, the area and population of which are shown in the margin. They are administered by Committees, each consisting of 9 members appointed by Government; they have charge of village roads and pounds within the Union, and receive fixed grants from the District Board for their maintenance.

Union.		Area in square miles.	Population.
Bāsudebpur	...	6	5,400
Itinda ...	...	10	13,450
Jadurhāti	...	10	11,922

MUNICIPALITIES.

The district contains no less than 26 municipalities, the inspection and control of which occupy much of the District Officer's time and energies. Their aggregate income in 1911-12 was 10½ lakhs (excluding the opening balances). The receipts have risen considerably of late years, mainly owing to the large sums received from Government, either as loans or grants, for carrying out drainage and water-supply schemes, and partly on account of revisions of the assessment and the imposition of fresh rates and taxes, such as latrine rates in the extended areas of Garden Reach and Tollygunge, lighting rates in Cossipur-Chitpur and Māniktāllah, and water rates in Tollygunge. The statement below gives at a glance the more salient facts relating to each municipality for the year 1911-12. More detailed information will be found in the articles dealing with each town in the last chapter:—

MUNICIPALITY.	Year of establishment.	Number of rate-payers.	Percentage to population	Number of Municipal Commissioners.	Incidence of taxation per head. Rs. A. P.
Bāduria	1869	1,936	14·1	12	0 6 7
Bārāset	1869	1,734	19·7	8	1 3 9
Barnagore	1869	3,976	15·3	19	1 11 9
Barrackpore, North ...	1869	1,916	11·8	15	0 14 2
Ditto, South ...	1869	1,765	14·8	2	0 15 0
Bāruipur	1869	1,152	21·4	19	1 5 7
Basirhāt	1869	2,504	13·6	15	0 6 7
Bhātpāra	1899	3,006	5·9	5	0 14 0
Budge-Budge ...	1900	2,093	11·6	9	1 15 1
Cossipur-Chitpur ...	1889	5,781	12·2	12	4 15 3
Dum-Dum, North ...	1870	904	17·9	9	1 4 10
Ditto, South ...	1876	1,776	13·7	9	1 4 1
Garden Reach ...	1897	3,956	8 8	12	2 13 2
Gārulia	1896	609	5·2	10	1 4 1
Gobardānga ...	1870	1,063	20·9	9	0 11 10
Hālishahar	1903	2,350	17·4	12	0 14 6
Jaynagar	1869	1,527	16·5	12	0 12 5
Kāmārhāti	1899	2,005	11·1	12	1 1 4
Maniktālā	1889	7,845	14·5	12	2 11 6
Naihāti	1869	5,400	29·6	9	1 0 8
Pānihāti	1900	2,435	21·9	10	1 1 7
Rājpur	1876	2,291	19·7	18	0 12 2
South Suburban ...	1869	6,302	22·4	10	1 0 2
Tāki	1869	959	18·4	9	0 7 2
Titāgarh	1895	1,614	3·5	10	0 15 9
Tollygunge	1901	3,270	17·5	9	1 8 7

The percentage of rate-payers in Titāgarh, Gārulia and Bhātpāra is the lowest in the Province, and the incidence of taxation in Cossipur-Chitpur is higher than in any other municipality, except Calcutta and the hill stations of Darjeeling and Kurseong.

The history of municipal government in the district is largely a record of the multiplication of municipalities in order to keep pace with the increase of population and the growing demand for better sanitation and conservancy, more particularly in the Suburbs of Calcutta and in the mill towns along the Hooghly. The application of the term "Suburbs of Calcutta" has varied widely at different periods. By Act XXI of 1857 the "Suburbs" were defined to include all lands within the general limits of the Panchānnagrām estate, and they were further defined under the Bengal Municipal Act of 1876. They then included the present municipalites of Cossipur-Chitpur, Māniktalā, Garden Reach, South Suburbs and Tollygunge, as well as so much of Calcutta as lies outside the limits of the "Old Town," which is bounded by Lower Circular Road and Tolly's Nullah. This unwieldy municipality, which was known as the Suburban Municipality, was split up into four parts in 1889. The "Added Area" and "Fringe Area Wards" were added to Calcutta, and the municipalities of Cossipur-Chitpur and Māniktalā were created, the name of the North Suburban Municipality being at the same time changed to Barnagore. These deductions still left the South Suburban Municipality of unmanageable size, and accordingly, in 1897, the Garden Reach Municipality, and, in 1901, the Tollygunge Municipality, were separated from it. The constitution of the present South Suburban Municipality, therefore, dates from 1901.

The following statement exhibits briefly the effect of these changes and also shows the changes which it has been necessary to make in the municipal government of the areas included in the Naihāti and the North and South Barrackpore Municipalities: the date entered after each municipality marks the year in which it was created.

North Suburban ...	Cossipur-Chitpur (1889). Barnagore (1889). Kamarhāti (1899).
South Suburban ...	South Suburban. Garden Reach (1897). Tollygunge (1901).
North Barrackpore ...	North Barrackpore. Gārulia (1896).
South Barrackpore ...	South Barrackpore. Pānihāti (1900). Titāgarh (1895).
Naihāti	Naihāti. Bhātpāra (1899). Hālishahar (1903).

Outside the suburban area the municipalities are still, to a greater or less extent, rural in character in that they include a certain quantity of cultivated land. As a rule, they consist of a central urban area with outlying villages, interspersed with paddy fields, that are grouped together for municipal purposes, but have little connection with each other or with the central area. The riverain tract, however, is, as stated in Chapter III, becoming increasingly populous, and in consequence increasingly urban, and the towns along it, more particularly the mill towns, are in a transition stage. For example, the question of accommodating the operatives who congregate in the neighbourhood of the mills, and the task of providing an adequate system of water-supply, drainage and sanitation for an industrial population are the chief problems of their municipal administration. Much has already been done to meet these wants, but only with the help of loans and grants from Government (supplemented in some cases by funds raised locally), for the municipal funds alone are inadequate to meet the expenditure necessary to provide a pure water-supply and an efficient drainage system. The following nine municipalities now possess a system of filtered water-supply :—Bhātpāra, Cossipur-Chitpur, South Dum-Dum, Garden Reach, Gārulia, Māniktalā, South Suburban, Titāgarh and Tollygunge, while the Naihati Municipality has taken up a scheme for the supply of filtered water from the Gauripur mills. The Bhātpāra, Gārulia and Titāgarh municipalities, it may be mentioned, were granted their supply from the local mills, free of cost, and, in introducing the system, incurred no expenditure except that of laying pipes and erecting hydrants. The following eight municipalities have undertaken comprehensive drainage schemes, and in most cases the work has been either partially or wholly completed :—Barnagore, Bhātpāra, Budge-Budge, Bārāset, Bāruipur, Basīrhāt, Gārulia and Titāgarh. The suburban municipalities of Cossipur-Chitpur and Māniktalā, and almost all the mill municipalities, have also done much to improve the sanitation of *bastis* within their respective areas.

In conclusion, the following remarks, illustrating the progress made by the municipalities in providing civic requirements, may be quoted from the Commissioner's last Quinquennial Administration Report :—" Most of the suburban and mill municipalities now provide an adequate supply of filtered water. In most of the municipalities there is an efficient system of conservancy, and the principal roads are lighted. In some of them drainage works have been started, and in others are under consideration. In no case, however, are funds sufficient to carry out expensive sanitary improvements independently and without any aid from Government. In fact, most of the important and costly sanitary schemes have only been undertaken with the help of loans, as well as of grants, supplemented in the case of the mill municipalites by substantial contributions from the mills. . . . The efficiency of municipal administration is not confined by any means to mill and riparian municipalities, where the Commissioners consist of Europeans and Indians, the former generally preponderating. There are many other municipalities which are extremely well managed entirely by Indian Commissioners, and I can certainly endorse the following remarks by the Magistrate of the 24-Parganas regarding these gentlemen :—'They show a spirit of independence tempered usually with good sense and a disposition to consider advice and act upon it unless they can show that it is bad.' "

CHAPTER XIV.

EDUCATION.

LITERACY. A fair indication of the extent to which education is diffused is afforded by the census statistics of literacy. The test of literacy is ability both to read and write, with this further qualification, that a person is recorded as literate only if he can write a letter to a friend and read the answer to it; all persons who are unable to do this are entered in the census schedules as illiterate. The total number of persons in the 24-Parganas who came up to this standard in 1911 was 300,818, representing 12 per cent. of the total population; the only other district of Bengal in which the people are more advanced is Howrah, where the ratio is 14 per cent. There has been a decided advance since 1901, the proportion of literate males having risen from 20 to 22 per cent. and of literate females from 13 to 17 per mille. The improvement is the more noticeable because the test of literacy was stricter than in 1901, when no conditions were laid down as to ability to read and write a letter.

Of the total number of literates, 222,203 are Hindus and 72,844 are Musalmāns, so that approximately there are three literate Hindus to every literate Musalmān. Taking the proportional figures, 25 per cent. of the Hindu males and 15 per cent. of the Musalmān males are literate, the corresponding ratios for females being 23 and 4 per mille respectively. There is a great disparity between the figures for males and females, for whereas 281,510 males can read and write, only 19,308 females can do so; in other words, the literate males outnumber the literate females in the proportion of 15 to 1. Education is nearly as general in the villages as in the towns, for 12 per cent. of the rural and 13 per cent. of the urban population are literate. Altogether 39,960 persons (38,528 males and 1,432 females) are able to read and write English, the ratio being 30 per mille in the case of males and 1 per mille in the case of females.

EDUCATIONAL STATISTICS.

The number of pupils under instruction increased from 52,000 in 1883-84 to 68,138 in 1892-93 and to 73,021 in 1900-01. In the next decade there was a further increase of 33,927 or 46 per cent., the returns of the Education Department for 1910-11 showing that there were 2,022 educational institutions with 106,948 scholars. Details of these schools and of their attendance are given in the marginal statement.

Schools.	Number.	Number of pupils.
High English ...	37	7,500
Middle " ...	62	5,559
Middle Vernacular	34	2,367
Upper Primary ...	131	8,490
Lower " ...	1,556	74,736
Technical ...	3	90
Training ...	10	114
Others ...	185	7,825
Private ...	4	267

According to the returns of the Education Department, the number of male scholars in 1910-11 represented 49 per cent. of the male population of school-going age, the corresponding proportion for female scholars being 7 per cent. The school-going age, it may be explained, is 5 to 15 years, and the number of children of this age is assumed to be equivalent to 15 per cent. of the population, but the census shows that in Bengal the actual proportion of children aged 5—15 is 27 per cent. for males and 25½ per cent. for females. The real percentage of school-going age is therefore much less than that shown in the departmental returns.

Thirty schools with 1,754 pupils are managed by Government, and three with 202 pupils by the District or Municipal Boards. The remainder are under private management, and, of these, 1,755 with 92,630 pupils receive grants-in-aid either from Government or from the District or Municipal Boards, while 230 with 12,105 pupils are unaided.

HIGH SCHOOLS.

The following is a list of the high schools in the district with the number of pupils on the rolls on 31st March 1913:—

Place.	Pupils.	Place.	Pupils.
Managed by Government.		*Aided*—contd.	
Bārāset	248	Baharu	222
Barrackpore	263	Barisa	291
Tāki	290	Barnagore (Victoria) ...	400
		Bāruipur	254
Aided.		Basīrhāt	481
		Behāla	173
Arbelia	135	Borāl	155
Ariādaha	281	Budge-Budge (P.K.) ...	270

Place.	Pupils.	Place.	Pupils.
Aided—concld.		*Unaided.*	
Diamond Harbour ...	280	Bawāli	288
Gobardānga	120	Dhānkuria	343
Gustia (K.N.)	140	Gārulia	148
Hālishahar	143	Kāmārhāti (Sāgar Dutt Free)	184
Harināvi (A.S.)	285	Madārhāt (Popular Academy)	140
Hatuganj (M. N. K.) ...	281	Majilpur	343
Jaynagar	393	Paikpāra (North Suburban) ...	167
Maheshtala	138		
Naihāti (Mahendra) ...	220		
Nārikeldānga (George) ...	295	*Girls' High English.*	
Nawābganj (Sridhar) ...	142		
Nibādhai	265	Barrackpore (Boarding) ...	86
Pānihāti (Tārānath) ...	195		
Sarisa	190		
Sodepur	126		
Vishnupur (Bishtupur) ...	181		

CHAPTER XV.

GAZETTEER.

Achipur.—Village on the Hooghly, situated 15 miles south-west of Alipore and 6 miles south-west of Budge-Budge, with which it is connected by the Orissa Trunk Road. The place derives its name from a Chinaman, referred to in old records as Atchew or Tong Achew, who was given a grant of land by Warren Hastings and started a sugar manufactory here. In a memorial which he submitted in 1781 to the Governor-General (Warren Hastings) and the members of the Supreme Council, he referred to the encouragement he had received by a grant of land, which he had cultivated with some success, and complained that his Chinese labourers were being enticed away by Chinese deserters from the ships in Calcutta. A notice was thereupon issued stating that Atchew was under the protection of Government, and that the Board wished to grant every encouragement to the colony of Chinese under his direction, and were determined to afford him every support and assistance in detecting and bringing to condign punishment any ill-disposed persons who inveigled away the Chinese labourers in his employ, who were under indentures to him for a term of years. Atchew died shortly after this, as appears from a letter, dated 8th December 1783, from the Attorney to the East India Company stating that he had applied to the executor of Tong Achew for the payment of a bond from the deceased to the Hon'ble Company.* On the 15th November 1804, we find an advertisement offering for sale "the estate of Atchepore, situated about 6 miles below Budge-Budge, with all the buildings, stills, sugar mills and other fixtures": the estate was said to consist of 650 bighas held by "pottah" from the Burdwān Rāj and paying rent of Rs. 45 per annum.†

Achipur at this time contained a powder magazine, at which vessels proceeding to Calcutta were required to deposit all gunpowder on board, except 100 pounds, which they were allowed to keep in order to fire salutes or signals of distress. When outward bound, they could take back the gunpowder so deposited.

* *Bengal, Past and Present,* Vol. III (Jan.-March 1909), pp. 137-8.

† *Ibid,* Vol. I (Oct. 1907), p. 204.

Severe penalties were prescribed for the breach of these regulations, which were laid down on the ground that "the explosion of a large quantity of gunpowder on board of ships lying off the town might be attended with the most destructive consequences to the town, to the inhabitants thereof and to the shipping in the Port."*

The village contains the grave of its founder, a characteristic horse-shoe shaped tomb. There is also a Chinese temple about a mile from the river. Its most noticeable features are a Chinese laver outside, a courtyard with walls covered with Chinese inscriptions, through which the shrine is approached, and a metal urn inside the latter, in which burning joss sticks are placed. The Chinese of Calcutta come here on pilgrimage every year about February, and use the temple for worship. There is also an abandoned and ruinous bungalow in the village, which was formerly occupied by the Collector of the 24-Parganas; some of the floors are paved with Chinese marble. An inspection bungalow of the Public Works Department is maintained here, and there is a post and telegraph office.

Alipore.—Headquarters of the district and a southern suburb of Calcutta. It is part of the district for judicial and revenue purposes, but its municipal administration is under the Corporation of Calcutta, and it is policed by the Calcutta police. It forms a ward (No. 23) of Calcutta, and in 1911 had a population of 19,749 persons, of whom 291 were Europeans. The population has increased by 11½ per cent. since 1901, mainly owing to the influx of newcomers to the Indian quarters, which has been stimulated by the extension of the electric tramway through it. It is also a popular place of residence for Europeans, and a number of new houses have sprung up recently, so that the old Penn estate has become a European colony. The growth of population would have been still greater had it not been for two opposing factors. In the first place *basti* lands have been acquired by Government, the Port Commissioners and the Calcutta Corporation, and large areas have been cleared either by them or by private parties. In the second place, acquisitions made by the Port Commissioners have practically depopulated the extensive area lying between the Boat Canal and the Tollygunge Circular Road on the one side, and between Diamond Harbour Road and Tolly's Nullah on the other. Partly on this account and partly on account of the large area occupied by Belvedere, the Zoological Gardens, and the gardens of the Agri-Horticultural

* *Bengal, Past and Present,* Vol. I (Oct. 1907), p. 153.

Society, Alipore is the most thinly-populated ward in Calcutta, there being only 16 persons per acre.

It contains the usual public offices of a district headquarters and a cantonment for native troops. The chief industrial concerns are the telegraph workshop, which in 1911 employed an average of 639 persons daily, and the Army Clothing Factory, in which there was a daily average of 380 operatives. The most interesting, and certainly the most imposing, building within its limits is Belvedere, which stands in extensive park-like grounds. Formerly a country house of Warren Hastings, it was purchased in 1854 for the residence of Sir Frederick Halliday, the first Lieutenant-Governor of Bengal, and was subsequently greatly improved and embellished by Sir Ashley Eden, Lieutenant-Governor of Bengal from 1877 to 1882. It continued to be the official residence of the Lieutenant-Governors till 1912, when that line of able administrators was ended by the appointment of a Governor, who took over Government House as his residence. Since then it has remained vacant, but it has recently been decided to keep in it the exhibits presented to the Victoria Memorial Hall, until the latter is erected. At or near the spot which is now the western entrance of Belvedere, on the Alipore Road, was fought the famous duel between Warren Hastings and Philip Francis, in which the latter was wounded. This duel is commemorated by Duel Lane, across the road, which leads to the Meteorological Observatory. No great distance away is Hastings House, the favourite residence of the great Governor-General, which was used as a guest-house for distinguished guests by the Government of India until the removal of the capital to Delhi. South of Belvedere lie the gardens of the Agri-Horticultural Society of India, founded in 1820 by Dr. Carey, the famous Baptist missionary, and north of it are the Zoological Gardens, which were opened by the then Prince of Wales in 1876.

Alipore Subdivision.—Sadar or headquarters subdivision, with an area of 1,164 square miles (of which 450 square miles are in the Sundarbans) and a population, according to the census of 1911, of 756,348 persons, the density being 650 per square mile. There are approximately two Hindus to every Musalmān, the actual numbers being 502,745 and 240,979, respectively. These figures exclude the Suburbs of Calcutta, *i.e.*, the three municipalities of Cossipur-Chitpur, Māniktala and Garden Reach, which actually form part of the subdivision, but are treated as a separate unit for census purposes; their area is 10 square miles and their population is 147,240, the mean density being 23 persons per acre.

The subdivision is bounded on the north by the river Hooghly, the city of Calcutta and the Bārāset subdivision, on the east by the Basīrhāt subdivision, on the south by the Diamond Harbour subdivision and the Bay of Bengal, and on the west by the river Hooghly. Taking the road from Alipore on the north to Jaynagar on the south as a dividing line, the western half is a low-lying tract with numerous marshes or *bils*. The eastern half is cut up by rivers, *khāls* and streams, which ensure better drainage, and there are fewer swamps, but in places the land is below high water level and the water is only kept out of the fields by high embankments. To the south there is a strip of the Sundarbans, about 50 miles long and 10 miles broad, which terminates in Bulcherry Island on the sea face. The greater portion of this Sundarbans country has been reclaimed and brought under cultivation, and consists of "lots," or blocks of land bounded by rivers and creeks and protected from inundation by embankments. Owing to its isolation, the population of this tract is sparse, and the southern extremity is thick jungle tenanted only by wild animals.

The Hooghly flows along the western boundary, and in the east of the subdivision the chief river is the Bidyādhari, which is connected with the Hooghly by Tolly's Nullah. This river has a circuitous course, for it flows through the Basīrhāt subdivision from east to west, turns south on entering this subdivision and then flows south-east, joining the Mātla river just above Canning. The Mātla is now only large enough for river steamers, but at one time was navigable by sea-going vessels as far as Canning. The Piāli is a subsidiary river, about 20 miles long, which connects the Bidyādhari with the Mātla. The Bhāngar Canal also connects the Bidyādhari at Kulti with the Baliāghāta Canal at Bāmanghāta, and is the principal route for boats coming from Khulna and the eastern districts to Calcutta.

The headquarters of the subdivision are at Alipore, and there are six towns, viz., Bāruipur, Budge-Budge, Jaynagar, Rājpur, South Suburbs and Tollygunge, each of which is dealt with in a separate article.

Bāduria.—Town in the Basīrhāt subdivision, situated on the right or west bank of the Ichāmai or Jamuna river. It is most easily reached from either the Arbālia or Gopmahāl stations of the Bārāsat-Basīrhāt Light Railway, being connected with both by second class roads. It is nine miles by river from Basīrhāt, from which place it can be visited in a steam launch. Maslandpur, a station on the Eastern Bengal State Railway, lies about 10 miles to the north, but the road is fit for travelling on horseback or in a bullock cart for only six months in the year, and in the rains is hardly passable.

Bāduria has a population, according to the census of 1911 of 13,680 persons, of whom 7,536 are Hindus and 6,142 are Musalmāns. It is the headquarters of a thana, and contains a dispensary, post and telegraph office, sub-registry office, and high school. The town forms a municipality, with an area of 12 square miles, which is divided into eight wards, viz., Bāduria, Arbālia, Tārāgunia, Magrāthi, Pura, Khorgāchi, Andarmānik, Māgurkhāli and Paddarpur. The municipal income is raised by a tax on persons at the rate of 12 annas per hundred rupees of income, Government and other public buildings being assessed at 7½ per cent. on their annual value. Latrine fees are also levied according to a prescribed scale. The water-supply is derived partly from tanks, but mainly from the river Ichāmati, which becomes brackish for a short time in the hot weather when the water is at its lowest. The line of drainage is from the river bank into the river and from the greater part of the town into a *bil* to the west. Five regular markets are held in the town limits at Bāduria, Arbalia, Pura, Tārāgunia and Paddarpur; the first three are held daily, that at Tārāgunia on Mondays and Thursdays and that at Paddarpur on Fridays and Sundays.

The chief *hāt* days at Bāduria are Tuesday and Friday. Sugar and molasses are manufactured in the town, and a considerable trade is carried on in those commodities and in jute, paddy and tobacco. An annual *mela*, the Bārwāri *Pūja*, which lasts three days, is held in May.

Bārāset.—Headquarters of the subdivision of the same name, situated 14 miles north-east of Calcutta on the Bārāset-Basīrhāt Light Railway and also on the Eastern Bengal State Railway, central section (from Dum-Dum Junction). The population of the town, at the census of 1911, was 8,790, of whom 5,017 were Hindus and 3,656 were Musalmāns. The population is not increasing appreciably, for the number returned in 1901 was 8,634, At the two previous censuses of 1891 and 1881 it was 9,754 and 10,533, respectively, but the decrease recorded in 1901 is probably to be attributed to the fact that in 1899 the area within municipal limits was reduced by the exclusion of some small outlying villages. The town contains the usual Government buildings found in a subdivisional headquarters, two munsifs' courts, a sub-registry office, a dispensary, a post and telegraph office, and a Government high school. The sub-jail is a there-storeyed building, popularly known as Vansittart Villa, and said to have been the country residence of Mr. Vansittart, a civil servant in the time of Warren Hastings.

Bārāset was constituted a municipality in 1869, and the area within municipal limits is 6½ square miles, divided between five wards, viz., North Bārāset, South Bārāset, Kāzipāra, Bāmanmura and Bādu. The municipal income is raised by means of a tax on persons at the rate of 13 annas per hundred rupees of income, Government and other public buildings being assessed at 7½ per cent. on their annual value. Latrine rates are levied at the rate of 7 per cent. on the annual value of holdings. There are three daily markets, one of which belongs to Government and is held in Bārāset, while the others are privately owned and are held in Kāzipāra and Bādu. The water-supply is obtained from tanks and a few wells; one large tank in the jail grounds is public property, and is reserved for the supply of drinking water. Wards Nos. 1 and 2 are drained by artificial drains, which lead to the fields on the south. The other wards are drained by the Sunthi Nadi, an old creek which passes southward to join the Bidyādhari river near Bhāngar.

Bārāset was formerly a place of greater importance than at present. In the early part of the nineteenth century it was the seat of a college for military cadets, which they entered on their arrival from Europe. On this account the town has been called "the Sandhurst of Bengal." From 1834 to 1861 it was the head-quarters of a district; one of the most distinguished of its District Officers was Sir Ashley Eden, Lieutenant-Governor of Bengal from 1887 to 1892, who was Magistrate here in 1859-60. Some remains dating back to the eighteenth century may still be seen about 4 miles to the north-east of the railway station. Here there is a large tank, called Madhumurali, which is said to have been excavated about 300 years ago by two merchant brothers named Madhu and Murali. To the north-west of the tank, a Mr. Louis Bonnaud, an indigo planter, is said to have had his factory, the site of which is marked by the remains of a large building overgrown with vegetation. On the bank of the tank at its south-eastern corner there is a high pillar, the object of which is unknown; and at its south-western corner there is an octagonal summer-house, now falling into ruin, the history of which is known. In his diary, Mackrabie (brother-in-law of Sir Philip Francis and Sheriff of Calcutta in 1775 at the time of Nuncomar's execution) speaking of some card-playing at "Barasutt" in February 1776 writes:—"Next morning such of us as were not too fatigued to leave our mattresses rode or walked to an octagon summer-house built upon an eminence by the late Mr. Lambert, who was the husband of Lady Hope. This is a pretty toy erected on an eminence and distant about a mile from Barasutt, with walks and flowering shrubs and gardens. The ashes of that gentleman (for his body was burned by his particular direction) are deposited under the building." Lady Hope was the widow of Sir William Hope, Bart., who perished in the Patna massacre in 1763, while she escaped to the Dutch factory. Her marriage to Mr. Lambert and the history of the latter are referred to as follows by the Revd. W. K. Firminger:—"On consulting the registers of St. John's (Calcutta), I found this entry against 27th April 1764—'William Lambert and Lady Margaret Hope, widow.' According to old custom, Lady Hope, although married to Mr. Lambert, elected still to be called by her higher social appellation, and on 1st February I find Mr. Lambert successfully petitioning for a passage for 'Lady Hope' on the *Lord Elgin.* Poor Lambert had apparently, like most Calcutta men of his time, many a painful loss of fortune. He entered the Company's service about 1760, and from 1763 to 1766 was Military Paymaster-General; then came a redistribution of offices, and our friend, married to a lady of consequence and blessed with a family, was suddenly left deprived of the bulk of his fortune, and also with a most inconvenient amount of unsold timber, which he had provided for the purpose of securing the new Fort William from river encroachments. At this time he seems to have fallen into discredit on the score of his accounts. I have traced him through several appointments. He was but a mere factor in October 1766; in January of 1767 he was clerk of the Court of Requests, and I have traced the name in the records of the 'Court of Cutchery.' In 1768, as we have seen, he sent his wife home. I can trace him fourth on the Murshidabad Council of Revenue in 1771, and I find his signature as a member of the Board of Inspection in the old Collectorate records of Chittagong. On 6th February he was appointed Chief at Dacca, but on the resignation of President Cartier he was given a place on the Council at Calcutta. He remained at Dacca till August, and then removed to Murshidābād, 'that place so much superior in point of climate': thence he came to Calcutta. In 1774 he became Chief at Dinājpur, and there he died on 18th September, 'at 8 A.M., after nine days' illness'. Surely this old octagon has a pathetic love story to tell us. Think of what Lady Hope must have passed through in the troubles of 1763, and of poor Lambert, after his long separation from his wife, asking nothing more than that his ashes might be buried beneath the dear old octagon at Bārāset."*

In conclusion it may be noted that in 1774 "the house and garden of Bārāset" were advertised for sale by auction as one of several houses belonging to the East India Company. The advertisement describes the property as follows:—"An upper-roomed house, part *pucca* and part *cutcha*, contains four bed-chambers, two back-stairs, two halls and verandah, a great stair-case to the south. A new *cutcha* detached building, consisting of a cook-room, bath-room, and bake-house with a stable at some distance from the house, built of posts and a straw chupper, very large and commodious, sufficient for four carriages and four and twenty horses. A garden surrounded with railing, and a ditch and a tank, and a very extensive avenue in front, which leads to the public road. Containing in all about 27 beeghas and 19 cottahs."†

At **Kāzipāra**, a suburb of the town, a large fair, which lasts three days, is held every year towards the end of December in honour of a Muhammadan Pīr or saint, called Pīr Ekdil Sāhib, of whom the following legend is told:—

There lived a king named Shāh Nīl, who was married to Ashik Nūrī, but had no children. One morning the female sweeper absented herself; and on being sent for, she refused to come before dinner, on the plea that by going early to Court she invariably had to see the faces of childless persons the first thing in the morning, which was an unlucky omen. The queen, struck by this remark, set out on a pilgrimage, in the hope that thereby she might obtain a child by the grace of God, and visited Mecca and other holy places. After thirty-six years of prayer an angel appeared to her, and having tried her faith in various ways promised her a child for two-and-a-half days. The queen returned home, and in due time gave birth to a son, which after two-and-a-half days was carried away by the angel, who took the shape of a fox. The child was brought up in the house of one Mullā Tar, and when he was about eight years of age came to Anarpur (the *pargana* containing Bārāset) riding on a tiger, which he could transform into a sheep at will. He crossed the Ganges on his stick, and came first to the village of Berua, where he planted his stick as a sign that he had entered into possession of the country assigned to him. The stick immediately grew into a thicket of bamboos. The boy then assumed the form

* *Bengal, Past and Present*, Volume II, No. 2 (October 1908), pp. 510-11. See also Volume I, pp. 202-3, and Volume II, pp. 261-2.

† *Notes on the History of Midnapore* (1876) by J. C. Price.

of a full-grown man, and proceeded to the house of one Chānd Khān, of Srī Krishnapur, a landholder of Anarpur, and begged a meal. Nur Khān, Chānd Khān's brother, refused to feed an able-bodied man, and told him to go and work at the mosque he was building. In proof of his supernatural powers, he lifted a block of stone of fifteen hundredweights up to the mosque, and miraculously prevented any bricks being laid on it. The mosque remained unfinished, and has furnished a proverb to the people, who call an incomplete undertaking a 'Chānd Khān's mosque.' After this, the stranger vanished. Again assuming the form of a boy, he called himself Dil Muhammad, and joined some cowherds. After working various miracles, he went to live with one Chhuti Miyān of Kāzipāra and tended his cattle. Sometimes he would ill-treat the cattle, and when the owners came out to punish him he transformed them into tigers and bears. On one occasion his cattle ate up a standing crop of paddy belonging to one Kumār Shāh, who complained to the headman of the village. An officer was accordingly sent to inquire into the matter, but he found the crops in this field better than any other in the neighbourhood. Upon his death a mosque was erected over his remains, and the fair is held at his tomb every year. About three hundred acres of rent-free land belong to the descendants of Chhuti Miyān for the service of the mosque.

Bārāset Subdivision.—Subdivision in the north of the district with an area of 275 square miles and a population, according to the census of 1911, of 292,791 persons, the density being no less than 1,065 per square mile. The Musalmāns predominate in the population, numbering 170,476, while the Hindus number 121,473. The subdivision is bounded on the north by the Rānāghāt subdivision of the Nadia district and the Bangāon subdivision of Jessore, on the east by the Basīrhāt subdivision, on the south by thana Hāroa of the Basīrhāt subdivision and thana Bhāngar of the Sadar subdivision, and on the west by the Barrackpore subdivision. There is no continuous natural boundary on any side except the south, where the Bidyādhari river forms the boundary line for a considerable distance. The subdivision is a little over 20 miles in length and 15 miles in breadth at the longest and broadest parts, and in shape resembles a square, except for a slight elongation in the south, which constitutes the Rājārhāt outpost. It forms an unbroken alluvial plain, studded closely with village sites enclosed by orchards or with scattered clumps of trees. The land slopes here and there into low swamps or *bils*, some of which cover a fairly large area. Except in the extreme south these *bils* are mostly cultivated with winter rice, but some of them are so much waterlogged as to be uncultivable, and in years of early and heavy rainfall may have to remain altogether uncultivated. In the south the *bils* are flooded with salt water coming from the Bidyādhari, which is a tidal river, or overflowing from the Dhāpa or Salt Water Lake, which is connected with them by several *khāls*. Such *bils*, though of little or no use to agriculture, constitute valuable fisheries. The country is traversed from the south-west to the north-east by the central section of the Eastern Bengal State Railway (to which the Calcutta-Jessore road runs parallel) and from west to east through its centre by the Bārāset-Basīrhāt Light Railway (opened in 1904). The only navigable water routes are the Bidyādhari river, and a branch of it that runs north past Baliāghāta, an important grain mart on the Bārāset-Basīrhāt Light Railway. There are several other rivers, such as the Jamuna (or Jabuna), Nawāi, Sunthi, Padda and Gobindakhāli, of which the Jamuna alone has a current throughout the year and the appearance of a river. The others have silted up and become dead channels. The beds of the Padda, which must once have been a large river, of the Gobindkhāli and the northern portions of the Nawāi and Sunthi have been 'bunded' up, *i.e.*, blocked by dams or embankments, and transformed into tanks or brought under cultivation. The subdivision contains two towns, viz., Bārāset, its headquarters, and Gobardānga.

Barnagore or Barānagar.—Town in the Barrackpore subdivision, situated on the Hooghly, 6 miles north of Calcutta, immediately north of Cossipur-Chitpur and south of Kāmārhāti. Its population in 1911 was 25,895, of whom 19,891 were Hindus and 5,792 were Musalmāns. It originally formed part of the North Suburban municipality, which was partitioned in 1889, part being constituted the Cossipur-Chitpur municipality, while the remainder had its name changed to Barnagore. Ten years later the northern portion of Barnagore was detached and formed into the Kāmārhāti municipality. The area of the municipality as now constituted is about 3½ square miles, and it is divided into four wards, viz.—(1) South Barnagore, (2) North Barnagore, (3) Dakhineswar and Ban Hugli, and (4) Pālpāra, Naopāra, Sainti and Nainān. The municipal income is raised by means of a rate on holdings at 6¼ per cent. on their annual value; latrine rates are also levied according to a prescribed scale. The water-supply is derived partly from the river Hooghly and partly from tanks and wells. There are over 1,000 tanks in the municipal area, of which only a fifth have wholesome water; one tank, in

Dakhineswar, is municipal property and is reserved for drinking water. The northern part of the town drains into the Dāntia Khāl, which forms part of the boundary between Barnagore and Kāmārhāti, and so passes into the Salt Lakes. The southern part drains eastwards into the paddy fields about Naopāra; and the river bank naturally, drains into the Hooghly. There is no public dispensary in the town, which is served by the North Suburban Hospital at Cossipur and the Sāgar Dutt Hospital at Kāmārhāti.

Barnagore is said to have been originally a Portuguese settlement, but it afterwards became the seat of a Dutch factory (the history of which is given in Chapter II), and during the greater part of the eighteenth century Dutch vessels anchored here on their way up to Chinsura. Old Dutch tiles of artistic design are still found in some of the buildings in the neighbourhood. Tieffentaller states that Barnagore was famous for its *bāfta* cloths, and Price in his *Observations* says that the cloth manufactories there determined Charnock to choose Calcutta as the site of the English settlement. The town was ceded to the British by the Duch Government in 1795, and the lands are comprised in the Barnagore Government estate, which is contiguous to the Panchānnāgram estate. The place used to be a favourite pleasure resort for European residents of Calcutta, but it is now a busy industrial centre and contains two of the largest jute mills on the Hooghly, while large quantities of oil are manufactured for export to Europe. The two mills in question are the Barnagore North and South Jute Mills, which in 1911 employed a daily average of 2,798 and 3,350 hands respectively: the Barnagore Branch Jute Mill, with 1,422 operatives, is in Baliaghāta. The town is the headquarters of a thana and has a bench of Honorary Magistrates and a high school (the Victoria High School). A *mela*, the Pānchu Charak *mela*, is held annually in April or May at Pālpāra. The name of the town is a corruption of Varāhanagar.

Barrackpore.—Headquarters of the subdivision of the same name, situated on the Hooghly, 14 miles north of Sealdah by the Eastern Bengal State Railway, and 15 miles from Government House, Calcutta, by road. The population in 1911 was 39,452. The town is comprised within two municipalities, viz., North Barrackpore, which has 11,847 inhabitants, and South Barrackpore with 27,605 inhabitants. The latter figure includes the population of the Barrackpore Cantonment, viz., 11,485. There is a considerable European community owing partly to the presence of British troops and partly to the fact that Barrackpore is a favourite place of residence for Europeans; altogether, 1,204 Christians were enumerated in South Barrackpore in 1911, of whom 839 were resident in the Cantonment. The military force stationed here consists of a battery of the Royal Field Artillery, a detachment of a British Infantry regiment (at present 4 companies) and a regiment of Native Infantry.

To the south of the Cantonment is Barrackpore Park, a large park, in which a golf course of 18 holes has been laid out. Within it are the tomb of Lady Canning, the wife of the Viceroy, who died of a fever contracted in the Tarāi when travelling down from Darjeeling, and Government House, the suburban residence of the Governor of Bengal and the country house of the Viceroys of India until the capital was removed to Delhi in 1912. The house appears to have been originally the residence of the Commander-in-Chief and was taken over by the Marquess Wellesley in virtue of his appointment as Captain-General and Commander-in-Chief of the forces in 1801. It is referred to as follows in Colonel Malleson's *Life of the Marquess Wellesley* (1889):—"Lord Wellesley had taken over, on his appointment as Captain-General, the residence heretofore allotted to the Commander-in-Chief. That residence was neither large enough nor commodious enough for the lodgment of the Governor-General of India and his suite. Yet it was desirable that one engaged in the arduous duty of governing India should possess a place in the country to which he could occasionally retire for rest and recreation. No locality appeared to the Marquess to be so well suited for such a purpose as the park at Barrackpore. It is the only piece of enclosed ground in India that has any resemblance to an English park. No sound from the outer world reaches the palatial residence. The majestic Hooghly flows calmly on one side, its surface gay with craft of varied shapes. On the other were magnificent trees, undulating grounds and a fine garden. Successive Governors-General have found there a place of real solace after the cares of Calcutta. The wife of one of the noblest of them, the courageous and high-minded Lady Canning, loved it so much that, when she died in India, her remains were transferred to the spot on which, when living, she delighted to sit and gaze at the river flowing beneath her. In this park Lord Wellesley designed to build a residence worthy of the representative of England's power in the East. He had the plans made and the estimates prepared. The builders were about to commence their work, when the Court of Directors, delighted to thwart him, forbade him to proceed. The work, in the style in which it was intended, was therefore abandoned."

The name of the town is due to the fact that troops have been stationed here since 1772. The Indian name for it is Chānak, which is sometimes said to be derived from the circumstance of Job Charnock having a country house here. There appears, however, to be no authority for this derivation, for the name dates back to a time anterior to Charnock. It may almost certainly be identified with the village of "Tsjannok" entered in Van den Broucke's map of 1660 and referred to by him as "the small town of Tsjannock," which his account shows was situated midway between "Cangnerre" and "Barrenger," *i.e.*, Kānkināra and Barnagore. Historically the place is interesting as the scene of two mutinies of the Bengal Army (in 1824 and 1857), which have already been described in Chapter II.

The South Barrackpore Municipality was constituted in 1869, but its area has been curtailed by the separation of the Titāgarh Municipality in 1895 and of the Pānihāti Municipality in 1900. Much of what is generally called Barrackpore, including the railway station, Government House and the Park, is comprised in Ward No. II (Chānak or Barrackpore) of this municipality. The municipal income is raised by means of a rate on holdings in Musalmānpāra Ward (at 5 per cent. of their annual value) and a tax on persons in the other wards; the latter is assessed at 12 annas on every hundred rupees of income, Government and other public buildings being assessed at 7 per cent. of their annual value. Latrine fees are also levied at 6 per cent. on the annual value of holdings. The municipal office is at Khardah, where the municipality maintains a dispensary for out-patients. An account of this village, which forms a separate ward, will be given later in this chapter. There is also a fine hospital within municipal limits, the Bholā Nāth Bose Hospital, which treats both in-patients and out-patients, and is maintained partly by endowments and partly by contributions from various public bodies. The chief educational institution is the Government high school; there is also a girls' high school.

The North Barrackpore Municipality was also constituted in 1869 and formerly included Gārulia, which was formed into a separate municipality in 1896. It has an area of 5½ square miles and is divided into ten wards, viz., Noāpāra, Ichāpur (two wards), Nawābganj (three wards), Palta, Dhitāra, Manirāmpur and Gānti. The municipal office is at Nawābganj, the residence of the Mandal family of zamīndārs; The Calcutta water-works are in the Palta ward of this municipality, which, however, derives no immediate benefit from them. There is a Government rifle factory at Ichāpur. The municipality maintains two dispensaries for out-patients, situated at Palta and Manirāmpur. At Nawābganj there is a high school, and two daily markets, one municipal and the other private. Two annual *melas* are held here, the Jhulan in August, which lasts 6 days, and the Gostāstamī in November, which lasts one day. Two main roads run through the municipality and are maintained by the District Board, viz., the Grand Trunk Road from Calcutta, which crosses the Hooghly at Palta, and the road to Kānchrāpāra, which takes off the Grand Trunk Road at Barrackpore.

The Cantonment is bounded on the south by the river Hooghly and the Barrackpore Park (in South Barrackpore), on the north and west by North Barrackpore and on the south by South Barrackpore. It derives its water-supply from the Palta water-works and is served by the Cantonment dispensary.

Barrackpore Subdivision.—Subdivision in the north-west of the district, with an area of 190 square miles and a population, according to the census of 1911, of 292,524 persons, the density being 1,540 per square mile. The population is increasing at a remarkably rapid rate owing to the development of the mill towns and the immigration of operatives, the growth since 1901 being no less than 42 per cent. The subdivision, which was formed in 1904 from portions of the Sadar and Bārāset subdivisions, is a narrow strip of land bounded on the west by the river Hooghly. The northern boundary is marked by the Bāgher Khāl, a creek flowing into the Hooghly, which separates it from the Nadia district. On the east lies the Bārāset subdivision, the boundary being marked for some distance by the Nawai Nadi, viz., from the Bariāti *bil* to the south-eastern extremity of the subdivision. On the south lie the municipalities of Cossipur-Chitpur and Māniktala, and the Salt Lakes, two square miles of the latter being included in the subdivision. There are no rivers of importance in the subdivision, but there are numerous *khāls* connecting the Hooghly with the low-lying country to the east. The land along the Hooghly is higher than that to the east, and the tendency is for the depressed inland basins to become more and more waterlogged as the *khāls* get silted up. The same tendency is noticeable in the case of the low-lying land to the south, which is connected by *khāls* with the Salt Lakes. The bank of the Hooghly is lined with mills, which provide employment for a large industrial population. There are twelve towns, all lying in the riverain strip of land, viz., proceeding from north to south, Hālishahar, Naihāti, Bhātpāra, Gārulia, North Barrackpore,

South Barrackpore, Titāgarh, Pānihāti, Kāmārhāti, Barnagore, North Dum-Dum and South Dum-Dum. There are cantonments at Dum-Dum and Barrackpore, a Government ammunition factory at Dum-Dum and a Government rifle factory at Ichāpur.

Bāruipur.—Town in the Sadar subdivision, situated on the banks of the Adi Ganga (an old channel of the Ganges now almost entirely silted up), 15 miles south of Calcutta, with which it is connected by the main line of the southern section of the Eastern Bengal State Railway and by the main road to Kulpi, which runs through the town. Its population in 1911 was 6,375, of whom 5,724 were Hindus. It is the headquarters of a thana and contains three Munsifs' Courts, a sub-registry office, a Bench of Honorary Magistrates, a high school, a dispensary for out-patients (maintained by the municipality), and a post and telegraph station. It is a mission station of the Society for the Propagation of the Gospel, which erected a large church here in 1846.

Bāruipur was constituted a municipality in 1869, the area within municipal limits being 2 square miles. There are six wards, viz., Bāruipur, Mandalpāra, Brāhmanpāra, Kāmārpāra, Bazar and Sāsan. The municipal income is raised by means of a tax of persons at the rate of Re. 1 per hundred rupees of income, Government and other public buildings being assessed at the rate of 7½ per cent. of their annual value. Latrine fees are also levied at the rate of 3 per cent. on the annual value of holdings. The water-supply is derived almost entirely from tanks. The town drains eastward into a *bil.* There is a daily bazar in the Bāruipur ward, and a *hāt* is held on Sundays and Wednesdays. Two *melās* are held every year, the Rās Jātra in November or December and the Rath Jātra in June or July. There are no local industries of any importance except *pān* cultivation and fruit-growing.

The name of the place is derived from the extensive cultivation of the former by the Bārui caste; in old records it appears under the corrupted form of Barrypore. Indigo was formerly manufactured, as appears from a statement in the *Gazette* of 16th January 1794, which sets forth—"We understand that the best indigo delivered on contract for the last year has been manufactured by Messrs. Wm. and Thos. Scott of Ghazipore and by Mr. Gwilt of Barrypore." In the early part of the 19th century it was the headquarters of the Salt Department in the 24-Parganas, and a Salt Agent and Medical Officer were stationed there. It was also the headquarters of a subdivision of the same name from 1858 to 1883, when it was amalgamated with the Sadar subdivision.

Basīrhāt.—Headquarters of the subdivision of the same name, situated on the right bank of the Ichāmati or Jamuna river in the north-east of the district. It is 27 miles from Bārāset, with which it is connected by the Bārāset-Basīrhāt Light Railway, 66 miles from Kidderpore *viâ* the Chitpur and Bhāngar Canals and 84 miles from Alipore *viâ* Canning. Its population in 1911 was 18,331, of whom 11,202 are Hindus and 7,129 are Musalmāns. It contains the usual public offices found at a subdivisional headquarters, two Munsif's Courts, a sub-registry office, dispensary and high school. The municipality covers an area of 10½ square miles, divided into 10 wards, viz., Dalchita, Nalkora, Dandirhāt, Sanipāla, Basīrhāt, Harishpur, Mirzāpur, Jārakpur, Tāntra and Bhābla. All obtain their water-supply partly from tanks and partly from the Jamuna or Ichāmati river, which becomes brackish in the hot weather. The greater part of the town drains into *bils* lying to the south, and the riverside into the river. There is an annual *melā*, the Bāruni *melā*, held in March or April, which lasts a week. Some *gur* and sugar are manufactured, but otherwise it has no industries of any importance.

There is one building of archæological interest in the town—the mosque known as the Sālik mosque. It consists of a hall measuring 36 feet by 24 feet, with two carved stone pillars, 8 feet high, supporting the roof; the latter has six domes arranged in two rows. The mosque is popularly reputed to have been built by one Alā-ud-din in the year 1305 A.D., but an inscription over the central *mihrāb* shows that it was erected by one Ulugh Majlis-i-Azam in 1466-67 A.D. The inscription is in Arabic, written in Tughra characters, and its translation is as follows:—"No God is there but He, and Muhammad is His Prophet. This mosque was built by the great and liberal Majlis, Ulugh Majlis-i-Azam—may his greatness be perpetuated—in the year 871."* An inscription on a moque at Pāndua in the Hooghly district shows that it was built by the same person in 1477 A.D. during the reign of Yusuf Shāh.†

Basīrhāt Subdivision.—Subdivision in the north-east of the district, with an area of 1,922 square miles and a population, according to the census of 1911, of 429,476 persons, the density being 223 per square mile. The smallness of the latter figure is

---

* *Pre-Mughal Mosques of Bengal*, by Monmohan Chakravarti, J.A.S.B., January 1910, p. 29.

† Hooghly District Gazetteer, p. 299.

due to the large area included in the Sundarbans, for the average is 1,267 in the Hāroa thana, 1,085 in the Basīrhāt thana and 904 in the Bāduria thana ; it falls, however, to 74 per square mile in the Hāsanābād thana, which is mainly Sundarbans country and, with an area of 1,620 square miles, accounts for more than four-fifths of the subdivision. There are ten Musalmāns to every eleven Hindus, the actual figures being 203,102 and 221,231 respectively.

The subdivision is bounded on the north by the district of Jessore, on the east by the district of Khulna, on the south by the Bay of Bengal and on the west by the Sadar and Bārāset subdivisions. The north and west consist of alluvial land, which is fairly well raised; on the south and east, where the delta is in a less advanced stage of growth, there is a network of tidal creeks winding their way to the sea through numerous islands and morasses. Altogether 1,584 square miles are included in the Sundarbans, the northern fringe of which has been reclaimed, or partially reclaimed, and is sparsely inhabited by cultivators whose huts may be seen dotted about the *ābāds*, as the reclaimed lands are called. The principal river is the Ichāmati or Jamuna. There are three towns, viz., Basīrhāt (the headquarters), Bāduria and Tāki.

Bhātpāra.—Town in the Barackpore subdivision, situated on the bank of the Hooghly, 22 miles north of Calcutta by the Eastern Bengal State Railway, the station being at Kānkināra The population of the town in 1911 was 50,414. It has grown extremely rapidly owing to the labour attracted by the mills, the figures for previous censuses being 21,540 in 1901, 14,135 in 1891 and 10,239 in 1881 ; in other words, the number of inhabitants has more than doubled in the last ten years and has increased nearly four-fold in 20, and five-fold in 30 years. Four towns only in Bengal, viz., Calcutta, Howrah, Dacca and Māniktala, have a larger population.

The town was formerly included in the Naihāti municipality, but in 1899 the portion lying south of Mukhtiāpur Khāl, which constituted two wards of Naihāti, were detached and formed, with the addition of Kānkināra, into the Bhātpāra municipality. The area within municipal limits (3 square miles) is about five miles long and half a mile broad, and is divided into three wards, viz., (1) Bhātpāra, (2) Mulājor, Athpur and Jagatdal, and (3) Kānkināra. The municipal income is raised by means of a rate on holdings assessed at 6 per cent. on their annual value ; latrine fees are also levied at the rate of 5 per cent. on the annual value of holdings.

Bhātpāra was formerly a seat of Sanskrit learning renowned for its *tols*. The *tols*, at which pupils are educated and fed free of cost, still exist, though in diminished numbers, and the pandits of Bhātpāra have a high repute as *gurus* and authorities on the Vedas. It is no longer, however, a quiet place in which a recluse can find a congenial home, but a busy industrial centre with several mills, situated chiefly in Kānkināra and Jagatdal. The marginal statement shows the mills at work and the average daily number of operatives employed in each in 1911, the aggregate being 27,160.

Place.	Mill.	No. of operatives.
Athpur ...	Auckland jute mill ...	2,494
Bhātpāra	Reliance ditto ...	3,117
Kānkināra	Titāgarh paper mill, No. 2.	1,256
	Kānkināra jute mill, No. 1.	3,683
	Ditto ditto, No. 2.	1,175
	Upper Anglo-India jute mill.	2,937
	Middle ditto do.	4,629
Jagatdal...	Lower ditto do.	2,685
	Fort Gloster jute mill ...	312
	Alliance ditto ...	3,002
	Alexander ditto ...	1,870

In Jagatdal the lines of two moats and two large tanks are reputed to be the remains of a fort erected by Pratāpāditya in the sixteenth century.

Budge-Budge.—Town in the Sadar subdivision, situated on the bank of the Hooghly. It is the terminus of a branch line of the Eastern Bengal State Railway, 16 miles long (from Sealdah), and is 10 miles from Alipore by road; the steamers of Messrs. Hoare, Miller & Co., plying between Armenian Ghāt and Uluberia, call here, the distance from Calcutta by river being about 15 miles. The population of the town in 1911 was 17,982.

Budge-Budge formerly contained a fort, which was captured by Clive in his advance on Calcutta in December 1756: an account of its capture will be found in Chapter II. The fort ceased to exist in 1793, as appears from the *Gazette* of 7th March in that year, which notified its abandonment and dismantling. Orders were issued that all the guns and stores were to be removed to Fort William, that the buildings and lands belonging to the East India Company were to be handed over to the Board of Revenue, and that all military expenditure on account of Budge-Budge was to cease. An advertisement also appears in the *Gazette* of the 23rd May, stating that these lands would be sold on 10th June. The only remains of the fort now visible are two moats, called the inner and outer moats. The former is still quite distinct, and contains water for most of the year, enclosing an island; the latter is partly distinguishable.

The town was constituted a municipality in 1900. The municipal income is raised by means of a rate on holdings assessed at 7½ per cent. on their annual value; latrine fees are also levied at the same rate. The drainage of the town is a difficult problem, as its site is low and water lodges throughout the rains. Most of the buildings are on artificially raised ground, and the pits from which the earth was dug to make their sites are filled with water and rank vegetation. There is a system of drainage by means of culverts and drains under the roads and railway embankment, which have sluices where they open into the river; but the land inside the embanked river face is so low that it cannot be satisfactorily drained without prohibitive expense.

Budge-Budge is the headquarters of a thana and has a Bench of Honorary Magistrates, a high school and a charitable dispensary maintained by the municipality. It is the oil depôt of Calcutta, at which ships laden with oil discharge. Several large firms have oil depôts here, the oil being pumped from tank steamers into huge circular iron tanks and distributed over the country by rail in tank trucks. The town also contains a cotton mill and two large jute mills. The former is the Empress of India cotton mill, which in 1911 employed on the average 677 hands daily; the latter are the Budge-Budge jute mill, in which the average was 6,942, and the Albion jute mill, in which the average was 3,251.

Canning.—A village, also known as Mātla, Canning Town and Port Canning, in the Sadar subdivision, situated on the Mātla river. It is the terminus of a branch of the Eastern Bengal State Railway, 28 miles south-east of Sealdah, and is 32 miles from Alipore by river, *viâ* Tolly's Nullah and the Bidyādhari. It may conveniently be visited in a steam-launch by the latter route as the first stage in a tour including Tāki, Basīrhāt and Bāduria. The village occupies a tongue of land, on the north of which flows the Bidyādhari. This river receives the waters of the Athārabānka and Karatoya, the united stream forming the Mātla, which flows past the south of the village on its way to the sea. The village is the headquarters of a thana known as the Mātla thana, and contains a sub-registry office, a post and telegraph office and a charitable dispensary.

The place is called after Lord Canning, during whose vice-royalty an attempt was made to establish a port here, though, according to Marshman, he "treated the whole project with supreme contempt." For many years before this fears had been entertained that the Hooghly was deteriorating, and in 1853 the Chamber of Commerce represented to Government the danger of the navigable channels closing and the necessity of establishing a port on the Mātla, which should be connected with Calcutta by a railway or canal. Lord Dalhousie's Government took the precaution of acquiring land for the proposed port, and in 1853 purchased lot No. 54 of the Sundarbans grants from the grantee for Rs. 11,000; this lot had an area of 25,000 *bighas*, or 8,000 acres, of which one-seventh was under cultivation, the remainder being uncleared jungle. In an adjoining lot, which lapsed to Government, an area of 650 acres was reserved for a site on which to build a town, and plans for laying it out were drawn up. In 1862 a municipality was constituted, and next year Government made over to the Municipal Commissioners its proprietary rights in the land, reserving to itself, however, the right to take up any land that might be required for public purposes, *e.g.*, for a railway station, public offices, etc. The cost of laying out and draining the town, constructing roads and protecting the river frontage was estimated at upwards of 20 lakhs, and in November 1863 the municipality opened a loan of 10 lakhs, upon debentures at 5½ per cent. interest repayable in five years. Only Rs. 2,65,000, however, were subscribed by the public. In 1864, a year of speculative mania, Mr. Ferdinand Schiller, of the firm of Borradaile, Schiller & Co., who was Vice-Chairman of the municipality, proposed to form a company, to be known as the Port Canning Land Investment, Reclamation and Dock Company, which would develop the port, construct docks, tramways, etc., and offered to subscribe 2½ lakhs to the municipal debenture loan in return for certain concessions.*

The proposal having been accepted by Government, the company received the gift in freehold of 100 acres of ground in the centre of the town, and also the exclusive right for 50 years of constructing tramways, wharves, and jetties, and of levying tolls in connection therewith. At the same time the company was required to excavate within two years a dock for country boats,

---

* Mr. Schiller also proposed, *inter alia*, that the Sundarbans should be formed into a separate district with headquarters at Canning, that the Magistrate of Bāruipur should be removed there, and that the Commissioner in the Sundarbans should make it his headquarters; also that a certain number of Government vessels and a certain quantity of the Government stores consigned to Calcutta should be sent *viâ* Canning. These proposals, however, met with little favour from Government.

and undertook the conservancy of the river bank. In return for these concessions the Municipal Commissioners obtained an immediate subscription of 2½ lakhs to the municipal loan, and had the prospect of sharing in the profits accruing from the works when the returns exceeded 10 per cent. on the capital invested. In spite of this addition to the funds, it was soon found that the sums raised from the public and the Port Canning Company were not sufficient for the works in hand. The municipality thereupon applied for a loan of 4½ lakhs, which Government granted in 1866, on certain securities, on the ground of the public and mercantile community having subscribed over 60 lakhs of rupees to the company for similar purposes.

The company started operations vigorously, laying down light-ships, moorings, buoys, etc. Its shares, which were issued in 1865, rose to a high figure, but they fell as rapidly as they rose, for it was soon realized that the sanguine expectations of the promoters were not likely to be fulfilled. In 1865-66 the port was visited by 26 ships, but five years later not a single ship put in there, and in the previous two years it was visited by only two ships, of which one was driven in by stress of weather. The failure of the scheme was patent. The Hooghly channels had not deteriorated as was apprehended, and trade showed no tendency to go to Canning. The company and municipality had come to loggerheads and were engaged in litigation, and the finances of the municipality were exhausted. No funds were available to meet the debentures that had fallen due, and Government refused to make any further advances. Finally, in 1871, the port was officially closed and the moorings taken up, while Government attached the property of the municipality under a Civil Court decree and placed it under the Collector of the 24-Parganas as a Government estate, which it still is. A few years later, the India General Steam Navigation Company had their vessels on the eastern river route loaded and discharged at Canning for about a year, and it was at one time proposed that the bulk oil depôt of Calcutta should be established there, but eventually Budge-Budge was selected for the purpose. The Port Canning Land Investment, Reclamation and Dock Company went into liquidation in 1870, and was reconstructed as the Port Canning Land Company, which is still in existence. It is under Pārsi management, the offices being at Bombay, and is chiefly engaged in zamīndārī, leasing out reclaimed land in the Sundarbans.

During its brief existence the municipality received and expanded upwards of ten lakhs, of which 8½ lakhs were obtained from loans. Government also disbursed either directly or through the municipality nearly 20 lakhs, besides making a railway for the benefit of the port, at a cost of over 60 lakhs, which failed to cover its working expenses. At Canning itself five jetties were built on the Mātla opposite what was called "Canning Strand," and two more on the Bidyādhari. A tramway was also laid down, and a wet dock, a rice mill capable of husking 90,000 tons of rice a year, a graving dock, goods sheds and landing wharves were constructed. Practically all that now remains is the railway, which has a certain amount of traffic in timber and other produce from the Sundarbans, some ruined jetties and the remains of a tramway line.

Cossipur-Chitpur.—Town in the Sadar subdivision situated on the bank of the Hooghly immediately north of Calcutta. It is bounded on the west by the Hooghly, on the south by the Chitpur Canal, which separates it from Calcutta and Māniktala, on the east by the Eastern Bengal State Railway line and on the north by Barnagore. The population in 1911 was 48,178, of whom 34,432 were Hindus and 13,037 were Musalmāns. The increase in the number of inhabitants since 1901 is 7,428, or 18 per cent., and is nearly entirely due to immigration, for the number of immigrants, *i.e.*, persons born outside the 24-Parganas, is 5,916 more than it was in that year and now accounts for two-thirds of the population. The growth of population would have been even greater if the census had been taken early in February instead of on 10th March. The majority of the operatives and labourers employed in the jute presses and factories come from up-country and reside in the town for about eight months in the year, three-fourths of them returning to their homes as soon as the jute season is over; their exodus generally commences in the end of February and terminates in March. The jute season in 1910-11 was dull and short, so that fewer labourers came, and those that did come left earlier than usual. Besides this, the Eastern Bengal State Railway has acquired within recent years extensive areas, containing thickly peopled *bastis*, which have been demolished; during the three years preceding the census some large jute presses were also erected on land that was formerly occupied by *bastis*. The inhabitants of these *bastis* could not all find accommodation within the town and had to move outside its limits. The average density of population is 23 per acre, but varies considerably in different wards, being 12 in ward No. 3, 23 in ward No. 4, 29 in ward No. 2 and 39 in ward No. 1. The latter ward, which lies along the Hooghly in the south-west of the town, contains a number of jute presses and factories, and is practically a part of Calcutta.

The population being largely composed of immigrant male labourers, who leave their families at home, there is a great disparity of the sexes, there being only 565 females to every 1,000 males.

The town was formerly part of the South Suburban municipality, but was constituted a separate municipality in 1889. The area within municipal limits is a little under 3½ square miles and is divided into four wards, viz., (1) Chitpur, (2) Cossipur, (3) Sainthi and (4) Belgāchia (including Tāla and Paikpāra). The municipal income is raised by means of a rate on holdings assessed at 7½ per cent. on their annual value. A water rate and lighting rate are also levied at the rate of 3 per cent. on the annual value of holdings, and latrine fees are assessed according to a prescribed scale. The town is supplied with filtered drinking water, which is obtained from the water-works of the Calcutta Corporation and distributed through the streets and to the houses by hydrants and pipes. The drainage of a small portion of the town along the river bank passes into the Hooghly, and the greater part drains eastwards into the cuttings of the railway embankments and thence through culverts into the Salt Lakes. There are two dispensaries, viz., the North Suburban hospital in Cossipur, a large institution which treats both in-patients and out-patients, and the Chitpur municipal dispensary, which treats out-patients only. There are two large private markets, called the Bheritala and Bibibazar markets, and two annual *melās* are held, viz., (1) the Mohan *melā* held at Phulbāgān on the Barrackpore Grand Trunk Road, which takes place at the end of December and beginning of January, and lasts five days, and (2) the Rām Lila *melā*, held in the grounds of a private garden house on the Grand Trunk Road, which lasts three days. The Belgāchia Veterinary College lies within municipal limits, and, in addition to the Government Gun Foundry and Shell Factory, there are a number of jute presses, sugar and other factories, which make the town a busy industrial place. The marginal statement show

Place.	Factory.	No. of operatives.
Chitpur ..	Strand Bank Jute Prss ...	500
	Union Jute Press ...	425
	Chitpur Hydraulic Jute Press	410
	Victoria do. do. ...	325
	Calcutta do. do. ...	320
	Ashcroft Jute Press ...	320
	Hooghly Hydraulic Jute Press	270
	Ocean Jute Press ...	218
Cossipur...	Gun and Shell Factory ...	1,271
	Cossipur Jute Warehouse ...	2,300
	Cossipur Sugar Works ...	739
	Bengal Hydraulic Jute Press	471
	Camperdown Jute Press ...	450
	Shellac Factory ...	420
	Jheel Jute Press ...	380

the principal factories and the average daily number of persons employed in each during 1911.

Dakshin Bārāset.—A village in the Jaynagar thana of the Sadar subdivision, situated on the bank of the Adi Ganga, an old channel of the Ganges. It is 27 miles south of Calcutta, with which it is connected by a metalled road, and 5 miles from the Magrā Hāt station on the Eastern Bengal State Railway. Including the subordinate villages or hamlets of Makundapur, Khātsāra, Pāirānai, Birttibāti, Māstikuri, Kālikāpur, Baliādānga, Nurullāpur, Ramākāntabāri, Abdulkarīmpur and Bāneswarpur, it has an area of about 4 square miles and a population, according to the census of 1911, of 6,507 persons. The name Bārāset is a British corruption of Bārāsat, which is said to be derived from the fact that the merchant Srīmanta, while journeying through *durjay Magra, i.e.*, the impassable Magra, worshipped Sata Bārās, or a hundred deities, to ensure his deliverance from danger. The tradition recalls the days when the Adi Ganga was a navigable river leading to the uninhabited Sundarbans and the perils of the Bay of Bengal. The prefix Dakshin was added to distinguish it from the town of the same name, which is the headquarters of the Bārāset subdivision.

The village contains a temple of Kālī and a shrine dedicated to the god Adyamahesh, regarding the foundation of which the following legend is related. One day, over 100 years ago, the cows of a Brāhman which were grazing in the fields were found to be yielding milk without their udders being touched. Next night the god Adyamahesh appeared to the head of the Chaudhuri family, who were the zamīndārs of the place, informed him that his image lay below the spot where this miraculous event had occurred, and directed him to build a temple there, of which the priests should be drawn from the family of Brāhmans to whom the cows belonged.

The village contains a post office, a Middle English School and a girls' school. Two bazars are held daily, one in the morning in front of the temple of Adyamahesh, and the other in the evening in front of the temple of Kālī. There are three annual *melās*:—(1) on the last day of Chaitra in the bazar near Kālī's temple, (2) on the first day of Baisākh in the Adyamahesh bazar, and (3) in April in honour of a *sannyāsi* named Achalānanda Tīrtha Swāmi, who was the *guru* or spiritual preceptor of Rājā Surendra Nārāyan Deb Bahādur, of the Sovābazar family of Calcutta. The tomb, or *samādhi*, of the Swāmi is in the village.

Diamond Harbour.—Headquarters of the subdivision of the same name, situated on the east bank of the Hooghly, which is here joined by the Diamond Harbour Khāl. The local name of the place is Hājīpur, and the creek is also known as Hājipur Khāl. It is 49 miles south of Calcutta by river, 32 miles by road and 38 miles by rail, and is connected with that city by a *pucca* road and also by a branch line of the Eastern Bengal State Railway, of which it is the terminus. Steamers also run across the Hooghly to Geonkhāli and Tamluk in the Midnapore district, and it is a stopping place of the Assam-Sundarbans steamers. The village, for it is nothing more, is about a mile in length and lies on both sides of the *khāl*. It contains the usual public buildings found in a subdivisional headquarters, four Munsifs' Courts, a sub-registry office, a post and telegraph office, a charitable dispensary, and a high school. In addition to the Subdivisional Officer, another Deputy Magistrate, a Sub-Deputy Magistrate, an Assistant Engineer of the Public Works Department and an Assistant Surgeon are stationed here. A harbour master and customs establishment are also maintained here to board vessels proceeding up the Hooghly, and the movements of shipping up and down the river are telegraphed to Calcutta and published, at intervals throughout the day, in the *Calcutta Telegraph Gazette*. It is further the local headquarters of the Salt Revenue Department, and a quarantine station has been opened for the accommodation of pilgrims returning from Mecca. The water-supply is obtained chiefly from a tank, the property of Government, which is reserved for the purpose.

About half a mile to the south is Chingrikhāli Fort, where heavy guns are mounted and the Artillery encamp annually for gun practice; the cantonment commences immediately to the south of the Subdivisional Officer's house.

Diamond Harbour was a favourite anchorage for ships a century and more ago. According to Hamilton's *East India Gazetteer* of 1815, "At Diamond Harbour the Company's ships usually unload their outward, and receive the greater part of their homeward bound cargoes, from whence they proceed to Saugor roads, where the remainder is taken in. There are mooring chains laid down here, and on shore the Company have warehouses for ships' stores, rigging, etc., and at an adjacent village provisions and refreshments are purchased." Graves dating back to this period may be seen in an old European cemetery situated beyond the telegraph station. Here "the clump of lofty casuarina trees, through whose foliage the summer wind whispers the music of the ocean, will indicate to those who pass by in ships the place where lie so many of our race, whose expectations of reaching their native land were at Diamond

Harbour thwarted by the call to a far longer journey."* The inscriptions on the graves date back over a century, the earliest being of the latter part of the eighteenth century. One epitaph records the death in 1832 at "Hidgelli Contai" (Hijili in the Midnapore district) of two young girls named Donnithorne within two days of one another; it also mentions the death of their mother but the last lines of the epitaph are missing. From the *Bengal Obituary*, in which the whole epitaph is given, we learn that she was the wife of a member of the Bengal Civil Service and died at Calcutta of a broken heart less than three months after her daughters.

In the compound of the house of the Assistant Engineer (which a mark in the floor of the verandah shows was erected in 1882 for the Trigonometrical Survey), there are two graves, of which one has an inscription to the memory of John Aitken, Inspector of Police, who, with his wife and child, was killed in the cyclone of 1864; the other has a slab but no inscription. Diamond Harbour suffered severely from this cyclone, which swept away the majority of its inhabitants; the loss of life within a mile of the river bank was estimated at four-fifths of the population. A mark on the wall of the Subdivisional Officer's cutcherry, which is fully 12 feet above the ground, marks the highest flood level.

Diamond Harbour Subdivision.—Subdivision in the south-west of the district, with an area of 1,283 square miles, of which 907 square miles are in the Sundarbans. The population in 1911 was 515,725, and the mean density 402 per square mile. The average is largely reduced by the Mathurāpur thana, which stretches into the Sundarbans and extends over no less than 966 square miles. In this thana the density is only 91 per square mile, but in all the other thanas the figure rises to over 1,100, reaching the maximum of 1,553 in Kulpi.

The subdivision is bounded on the west by the Hooghly, which separates it from the Midnapore district, on the north and east by the Sadar subdivision, and on the south by the Bay of Bengal. Along the sea face there are several islands, including Sāgar Island and Fraserganj. Two others are called the Tengra Chars (one old and the other new), and a third has formed in the bed of the Muriganga (also called the Bārātala or Channel Creek) to the east of Sāgar Island; an attempt has been made to bring the old Tengra Char under cultivation, but the other two are still to a large extent below high water level during spring tides. The land consists almost entirely of a series of low-lying basins

* *Bengal, Past and Present*, Vol. III, No. 1 (Jan.-March 1909), p. 159.

sloping inwards from the river banks, which are apt to be inundated whenever there is excessive rainfall, such as occurred in September 1900, when the country resembled an inland sea with the villages rising, like islands, above the waste of water. The country north of the Sundarbans is enclosed, on the west, south and east, by the Hooghly embankment, in which there are sluices for purposes of drainage, such as the Sātpukur sluice in the Mathurāpur thana, the Tengra and Kulpi sluices in the Kulpi thana, and the Bendal sluice in the Diamond Harbour thana. In the Sundarbans area private proprietors, or *lotdārs*, as they are called locally, have erected embankments round their lots, which protect the tracts under cultivation from the ingress of salt water. Lots Nos. 1-27 and 110-116 lie within the subdivision, and are now almost entirely under cultivation. Cultivation, in fact, extends to the south-west of the Sundarbans with the exception of an area of protected forest towards the southern extremity. To the south of the cultivated area the land is still covered with a dense low scrubwood, above which isolated forest trees raise their heads here and there. The chief navigable river is the Hooghly, which is joined by several feeder creeks or *khāls*, viz., (1) the Kātākhāli, a stream that is now nearly silted up, which debouches 3 miles north of Falta, (2) the Balarāmpur, which also joins the Hooghly near the Falta Fort, (3) the Nila Khāl, Kholākhāli, Hāra and Diamond Harbour Creek, all in the Diamond Harbour thana, of which the Hāra is silted up in all but a few places, and (4) the Kulpi Khāl and Tengra Khāl in the Kulpi thana. The principal channels in the interior of the subdivision are the Magrā Hāt Khāl, Kāorāpukur, Lakshmīkāntapur Sangrāmpur, Sātpukhur, Bānstala, Ghughudānga, Gundākāta and Andarmānik, of which the two first mentioned are the principal trade routes. The Magrā Hāt Khāl connects Magrā Hāt, Jaynagar and Surjapur; the Kāorāpukur affords communication between Magrā Hāt, Nainān and Jhinki, and all rice-laden *dongas* proceed by it to Chetla.

Dum-Dum.—Town in the Barrackpore subdivision, situated 7 miles north-east of Calcutta by rail and 8 miles by the Jessore road. The railway station forms the junction of the eastern and central sections of the Eastern Bengal State Railway. At Dum-Dum Cantonment there is another station on the latter section. The town is divided between the two municipalities of North Dum-Dum and South Dum-Dum, which have a population, according to the census of 1911, of 8,865 and 12,874, respectively; the figure for North Dum-Dum includes the population of the cantonment, viz., 3,818.

The name Dum-Dum is a corruption of *Damdama*, meaning a raised mound or battery. It appears to have been first applied to an old house standing on a raised mound,* of which the following account is given by Mr. R. C. Sterndale in the Annual Report of the Presidency Volunteer Reserve Battalion for 1891, in which year it was used as the Volunteer headquarters—"Dum-Dum House, or, as it is sometimes called by the natives, the Kila (the fort), is a building of some historic interest." It is probably one of the oldest existing buildings in Bengal, as it was in existence, though not in its present form, before the sack of Calcutta by the Nawāb Sirāj-ud-daula in 1756. The first mention of it occurs in Orme's *History of the War in Bengal.* He states that when Clive marched through the Nawāb's camp at Sealdah, on the morning of the 8th Feburary 1757, in a dense fog, he crossed the Dum-Dum Road. "This road," says the historian, "leads to Dum-Dum, an old building stationed on a mound." The cantonment and station of Dum-Dum were not established until nearly fifty years later, but the Bengal Artillery used to come out to Dum-Dum to practise on the plain, when the officers used to occupy the old building, while the men were camped in the grounds.

"The building appears to have been originally a one-storeyed blockhouse, so constructed as to secure a flank fire along each face, with underground chambers or cellars. The walls were of great thickness, from 4 to 8 feet thick, while they were further strengthened by massive buttresses, between which the walls were apparently loopholed for musketry. No authentic account of the origin of this building can be found, but it was probably either a Dutch or Portuguese factory. The native tradition is that the mound on which it stands was thrown up by a spirit in a single night, and to this day the house and grounds have the reputation of being haunted. Some time after the battle of Plassey, Lord Clive made the old building his country-house, altering the lower storey, so as to destroy its character as a defensive position, and building a fine upper storey; the grounds were also laid out with great expense and taste in the then prevailing formal Dutch style. Bishop Heber, nearly seventy years ago, speaks of this house as then presenting a venerable appearance and being surrounded by very pretty walks and shrubberies. No remains of these now exist, though the lines of the old walks and garden paths may be traced through the thin turf in the dry summer. From its elevated position and the massiveness of its

* An old house standing on a hillock in the Fort at Monghyr (demolished a few years ago), which dated back to Mughal times, was also called Damdama Kothi, *i.e.*, the Dum-Dum House.

structure, the old house would be still capable of a stout defence against anything but artillery."

At Dum-Dum, on the 6th February 1757, was concluded the treaty by which the Nawāb of Bengal ratified all privileges previously enjoyed by the English, made restitution of Calcutta, Cossimbāzār and Dacca, permitted Calcutta to be fortified and granted freedom of trade and liberty to establish a mint. A cantonment was established in 1783, previous to which the place had been the practice ground of the artillery,* and it was the headquarters of the Bengal Artillery until 1853, when they were removed to Meerut. In the latter part of the eighteenth century Dum-Dum was a fashionable place of resort for the European residents of Calcutta. "As Dum-Dum grew," writes Miss Blechynden in *Calcutta, Past and Present*, "it became the fashionable resort for Calcutta society, and many a gay cavalcade of fine ladies and gentlemen passed along the raised Dum-Dum road to be present at a grand review. The gay dames and gallants have long slept in their scattered tombs, but the memory of their passing to and fro still lingers in the countryside, where the simple village folk, as they gazed after them across the level expanse of their rice fields, threaded their own exclamations of pleasure at the sight on the melody of a song, which may yet be heard when, in the quiet evening hour, mothers croon their babes to rest—

Dekho meri jān!
Kampāni nishān!
Bibi gia Dum-dumma,
Oori hai nishān.
Burra sāhib, chota sāhib,
Bānka Kāpitān,
Dekho meri jān,
Lia hai nishān."

"Which may be freely translated—

"See, oh! life of mine!
The Company's ensign.
The lady to Dum-Dum hath gone,
Flieth the ensign.
Great men, little men,
Officers so fine,
See, oh! life of mine!
Goeth the ensign."

---

* Colonel Pearse in a letter, dated 23rd February 1775, says that his corps was encamped at tents in Dum-Dum to carry on practice, which usually lasted two months.

From 1861 to 1893 Dum-Dum was the headquarters of a separate subdivision, which was amalgamted with the Bārāset subdivision in the latter year. Since then the civil and criminal administration of the cantonment has been vested in a Cantonment Magistrate, who is also Cantonment Magistrate of Barrackpore.

The Dum-Dum cantonment has an area of 1½ square miles, and is bounded on the north, west and south by the South Dum-Dum municipatity; to the east lie rural tracts forming part of the Dum-Dum thana. There is a fine range of barracks, built round a square, which are occupied by a detachment of a British infantry regiment. It also contains a Protestant church (St. Stephen's), capable of containing seven to eight hundred people, a Roman Catholic church and Wesleyan chapel, a European and Ntaive hospital, a large bazar, and several large clear-water tanks. Within the balustrade, which surrounds the Protestant church, is a handsome pillar of the Corinthian order raised, by his brother officers, to the memory of Colonel Pearse, the first commandant of the Artillery regiment, who died in Calcutta, 15th June 1790. In front of the mess-house there was another monumental column raised to the memory of the officers and men who fell during the insurrection and retreat from Kābul in 1841, but more especially to Captain Nicholl and the officers and men of the 1st Troop, 1st Brigade, Horse Artillery, who were cut down to the last man in defence of their guns. This was blown down by a gale in 1852, and the pediment, with the marble slab containing the names of the officers and men, is all that now remains. The cantonment also contains the ammunition factory of the Ordnance Department, which manufactures arms, shells, etc., and has given its name to the "Dum-Dum bullet"; it employed a daily average of 2,681 operatives in 1911. Filtered water is supplied from the Calcutta water-works by pipes and hydrants.

The North Dum-Dum municipality has an area of about 5½ square miles, including the cantonment, and stretches for about 4 miles from east to west. In shape it resembles a dumb-bell, having two broad ends joined by a narrow neck. There are two wards, viz., Kādihāti and Nimta; a considerable portion is rural in character. The municipal income is raised by means of a tax on persons at the rate of 12 annas per hundred rupees of income; latrine fees are levied at 3 per cent. on the annual value of holdings. The water-supply is derived mainly from tanks, one of which, the Nimta Dighi in the north of Nimta, is of very large size. There is no efficient system of drainage. The rainfall of Ward No. I (Kādihāti) finds its way by *kutcha* drains to some extent into the Nawai Nadi, which finally flows into the Bidyādhari. Parts of Nimta and Birāti drain into the Nikauri Khāl, which leads into the fields to the south and there ends, for it is dammed up and its bed cultivated. Two small *melās* are held: one, called Chaukudhāni's *melā* is held in Nimta in February; the other, called Fakīr Sāhib's *melā*, is held in Gauripur in honour of a Muhammadan saint named Shāh Farīd. There is a municipal dispensary in Birāti.

The South Dum-Dum municipality has an area of about 5 square miles and is divided into three wards. The municipal income is raised by means of a rate on holdings in Ward No. 3, at 5 per cent. of their annual value, and a tax on persons in the other two wards, assessed at 1 per cent. on the annual income of the assessees. Latrine fees are levied according to a prescribed scale, and a water-rate at 2 to 4 per cent. on the annual value of holdings. Filtered water is obtained from the Calcutta water-supply system and distributed by hydrants. The greater part of the municipality drains into the Bāgjola Khāl, which enters it at the north, and passes away at the south-east, discharging into the Salt Water Lakes. An annual *melā* called the Sākurdi *melā* is held in Bāgjola in February, and lasts seven days. Two daily markets are held at Garbhānga and Nàgar Bazar for the sale of fish and vegetables, respectively. The municipality maintains a dispensary, which treats out-patients only. There is a large jute mill, within municipal limits, at Dakhindwāri. Three miles south of Dum-Dum is Pātipukur, a station on the Eastern Bengal State Railway, which is rising into importance as a terminus of the jute traffic.

Falta.—A village in the Diamond Harbour subdivision, situated on the bank of the Hooghly nearly opposite to the spot where it is joined by the Dāmodar. It is the headquarters of a thana and the site of a fort, mounting heavy guns, which forms one of the defences of the river Hooghly. In the eighteenth century the Dutch maintained a station here, to which the English retired after the capture of Calcutta by Sirāj-ud-daula in 1756, and at which they remained until a sufficient force had been collected for its recapture. Further details of their stay at Falta will be found in Chapter II.

The pages of the Gazette in the latter part of the eighteenth century and the beginning of the nineteenth century contain some advertisements relating to the place. The Gazette of the 15th April 1790 gave notice of the sale of the factory and grounds at "Pulta" and Fort Gloster with the powder-works at Manicolly; and on 8th July the result of the sale was announced as follows:—Pulta factory Rs. 5,800, Pulta bleaching ground Rs. 5,800, Old powder works Rs. 3,000, and Fort Gloster Rs. 2,450. The situation of the places mentioned makes it practically certain that Pulta is a misspelling of Fulta or Falta.

In the early part of the nineteenth century there was a large farm here under European management, as appears from three other advertisements. The first, which is dated 22nd July 1802, states that John Francis Gammidge admits John Saunders, late victualler to the East India Company, to partnership in the farm; in the second, which appeared in 1806, Gammidge and Saunders advertise lime juice, put up in kegs, far sale as an antiscorbutic. A third advertisement gave notice of the sale in 1815 of the Falta Farm and Tavern conducted by Messrs. Higginson and Baldwin.

Fraserganj.—Island in the extreme south of the Diamond Harbour subdivision. It is bounded on the north and west by the Pāttibunia Khāl, on the east by the Sattarmukhi river and Pukuriāber Khāl and on the south by the Bay of Bengal. It has an area of about 15 square miles, and is 9 miles long from north to south, its average breadth being 3 miles. It has a sandy beach facing the Bay, north of which are sand dunes, which a line of trees separates from the land to the north. There is one large fresh-water *jhil*, about a mile in length, lying between two of the dunes. The island is called locally Nārāyantala, and in the Admiralty charts bears the name of Mecklenberg Island. It was called Fraserganj after Sir Andrew Fraser, Lieutenant-Governor of Bengal from 1903 to 1908, during whose tenure of office a scheme of reclamation and colonization was undertaken and steps taken to develop the place as a health resort for the inhabitants of Calcutta and its neighbourhood. With this object about two-thirds of the island was cleared of jungle and roads and embankments were constructed. A dispensary and school were started, a post office opened, a golf course laid out and a dāk bungalow built on the highest dune. The work of reclamation, however, proved unexpectedly costly, and cultivators could not be induced to settle on the island on remunerative terms, so that the work was given up by Government. While it was in progress, a number of house sites were discovered with large tamarind trees and *mansa* trees (*Euphorbia nivulia*), growing near them, and in the south-east of the island four old kilns and scattered bricks were found—all proofs that the island was formerly inhabited.

Garden Reach.—Town in the Sadar subdivision, situated on the Hooghly, south-west of Calcutta, from which it is separated by the Nimakmahāl Ghāt road. Its population in 1911 was 45,295, of whom 21,286 were Hindus and 23,706 were Musalmāns. The mean density is 21 per acre, the average for each of the circles into which the municipality is divided being 22 in Circle No. 1, 19 in Circle No. 2, 13 in Circle No. 3 and 10 in Circle No. 4; the highest density is found in the circle next to Calcutta, and it diminishes the further one proceeds from the city. The increase of population since 1901 amounts to 17,084, which is partly the result of an extension of the municipal limits, the added area having a population of 6,444 or one-third of the net gain. It is also partly due to the immigration of mechanics and artisans attracted by the high wages obtainable in industrial works: at the time of the census 10,644 males, or nearly two-fifths of the male population, were employed in the jute and cotton mills and in the dockyards. These immigrants are, for the most part, unaccompanied by their families, and there is consequently a marked disparity of the sexes in the town, there being three males to every two females.

Garden Reach is a town of modern growth. In the eighteenth century it was the site of the small Mughal fort of Aligarh, opposite to which, on the other bank of the Hooghly, stood the fort of Tanna, both of which were taken by Clive in his operations for the recapture of Calcutta in 1756. In the latter half of that century the place became a fashionable suburb of Calcutta, and there are many fine houses along the river bank, mostly built between 1768 and 1780, which were the palaces of the Calcutta merchants. Its popularity declined when the King of Oudh settled at Mātiāburuz with a large entourage (after his deposition in 1856), and the better class of European residents, finding its amenities diminished, gradually deserted it in favour of Alipore and Ballygunge; some of the ex-king's descendants still reside in the locality. Most of the large houses are now used as the offices and residential quarters of the large factories and mills that have been established here.

Garden Reach was formerly part of the South Suburban municipality, and was formed into a separate municipality in 1897. The municipal income is raised by a rate on holdings, which is assessed at $7\frac{1}{2}$ per cent. on their annual value; latrine fees are also levied according to a prescribed scale, and a water-rate is assessed at $3\frac{1}{8}$ per cent. on the annual value of holdings.

The town is a busy industrial place, the principal concerns being (1) the Clive Jute Mills, of which mill No. 1 employed a daily average of 2,917 hands in 1911, and mill No. 2 of 2,775; (2) the Bengal Cotton Mills, with 2,000 hands; (3) the dockyards of the India General Steam Navigation Company and the Rivers Steam Navigation Company with an average of 1,331 and 1,271 respectively; (4) the Victoria Engineering Works of Messrs. John King and Company 167; and (5) the factory of the North-West Soap Company 183. The Lower Hooghly Jute Mill is situated at Badertala, but remained closed in 1911. Other buildings of interest are the King of Oudh's palace at Mātiāburuz (now dismantled), the Army Remount Depôt, the Emigration Agency for Trinidad, Fiji and Jamaica, the Emigration Agency for Demerara and Natal, the coal depôt and wharf of the British India Steam Navigation Company at Bracebridge Hall, and the Government (Public Works Department) brick-fields at Akra. A large trade in straw is carried on, the straw being brought in large country boats and landed at Bichāli Ghāt.

Gārulia.—Town in the Barrackpore subdivision, situated on the bank of the Hooghly, 1 mile from Shāmnagar station on the Eastern Bengal State Railway (19 miles north of Sealdah) and 5 miles by a *pucca* road from Barrackpore. Its population in 1911 was 11,580. The town was constituted a municipality in 1896, Wards Nos. 2 and 3 of North Barrackpore being separated to form it. It has an area of 2½ square miles and is divided into three wards, viz., (1) Naopāra, (2) Mills and Bazar and (3) Gārulia; the most populous ward is No. 2. The village of Shāmnagar, which forms the subject of a separate article in this chapter, is within municipal limits; this village contains the Shāmnagar Jute Mills and the Dunbar Cotton Mill. The income of the municipality is derived from a rate on holdings, which is assessed at 6 per cent. on their annual value; latrine fees are also assessed at 1 anna 3 pies a month per head (adult) of the population in the case of residential houses and at 2 annas a month per compartment in the case of cooly lines The municipality maintains an out-patient dispensary, and the two mills keep up well-equipped private dispensaries for their employés, who form a large proportion of the population. Filtered water is supplied to the town by the mills. The town contains an unaided high school.

Gobardānga.—Town in the Bārāset subdivision, situated on the east bank of the Jamuna or Ichāmati river, 36 miles north-east of Calcutta, with which it is connected by the central section of the Eastern Bengal State Railway. Its population in 1911 was 5,070. It constitutes a municipality, the area within municipal limits being 3 square miles. There are six wards, viz.,

(1) Gobardānga, Bābupāra and Gaipur, (2) Gobardānga, (3) Gobardānga, Kāntiākhāl and Sarkārpāra, (4) Khantura, (5) Haidādpur and Raghunāthpur and (6) Gauripur. The municipal income is raised by means of a tax on persons assessed at 12 annas per hundred rupees of income. There are an out-door dispensary maintained by the Mukherji family of zamīndārs, which was removed from Government supervision in 1889, a high school and a Bench of Honorary Magistrates. Trade is carried on in jute and molasses, and there are several sugar factories. Tradition points to this place as the spot where Krishna tended his flocks. Gaipur is said to have been the home of the *gopīnīs*, or milkmaids, with whom he sported; an embankment across the river is called Gopīnīpota, and an adjoining village bears the name of Kānhainātsāl, meaning "Krishna's pleasure-seat."

Hālishahar.—Town in the Barrackpore subdivision, situated on the bank of the Hooghly, 26 miles north of Calcutta, with which it is connected by the Eastern Bengal State Railway. Its population in 1911 was 13,423. It was formerly part of Naihāti, but was constituted a separate municipality in 1903. The municipal income is raised by a tax on persons, which is assessed at 10 annas per hundred rupees of income; latrine fees are also levied at the rate of 5 per cent. on the annual value of holdings. The municipality maintains a dispensary for out-door patients; there is also a high school and a Bench of Honorary Magistrates. Kānchrāpāra, a village containing the Locomotive and Carriage Works of the Eastern Bengal State Railway, is within municipal limits.

The place was formerly called Kumārhāta, and was a noted home of Pandits. Rām Prasād Sen, "that great saint and poet of the eighteenth century, whose name is known and revered throughout Bengal,"* was born here about 1720 A.D., and here composed many of his poems. When a young man, he was employed as a *sarkār* or accountant by a Calcutta merchant, but filled his ledgers with poems instead of with figures. When the head accountant reported his negligence to his master, the latter, admiring his talent, sent him back to Kumārhāta with a pension, and here, free from financial cares, he gave full vent to his poetical genius. His home is visited every year by a number of Hindus, and an annual *melā* is held in his honour.

Hāroa.—A village in the Basīrhāt subdivision, situated 6½ miles south of Deulia, with which it is connected by a *pucca*

* D. C. Sen, *History of Bengali Language and Literature* (1911), p. 656.

road. The village is said to derive its name from the fact that the bones (*hār*) of a Muhammadan saint, named Pīr Gorāchānd, were buried here; an annual fair is held in Phālgun (February) in honour of the saint, who is said to have lived six hundred years ago. Tradition states that this holy man came to Bālinda, the *pargana* within which Hāroa is situated, and settled on the banks of the Padma, close to the house of one Chandraketu, a rich landholder and staunch Hindu. Gorāchānd at once set to work to induce Chandraketu to embrace the faith of Islām. He performed several miracles before him, such as changing a piece of iron into a plantain, and causing a common fence to produce *champa* flowers. He also restored to life a Brāhman, who had been slain by the female monster Biroja. These miracles, however, did not shake Chandraketu's faith.

Unsuccessful in his attempt to convert Chandraketu, Gorāchānd next proceeded to Hāthiāgarh *pargana*, which was ruled over by Akhiānand and Bakānand, the sons of Rājā Mohidānand, who practised human sacrifice, offering up one of his tenants every year. On the occasion of Gorāchānd's visit, the lot for the next victim had fallen upon one Mūmīn, the only Muhammadan tenant of the Rājā. Gorāchānd offered to become the proxy of his co-religionist, but when taken before Bakānand he refused to fulfil his promise. A fight ensued, in which Bakānand was slain. Akhiānand, on hearing of his brother's death, invoked the aid of his guardian deity, Siva, who supplied him with a weapon, with which he took the field against Gorāchānd and severely wounded him. Gorāchānd asked his servant to procure some betel-leaves to apply to the wound, but he could not obtain any. Hence it is said that betel-leaf can never grow in Hāthiāgarh, and it is so far true that none is cultivated there. Gorāchānd returned wounded to Kulti Behārī, about 4 miles from Hāroa, where he was abandoned by his servant, who gave him up for dead. It is said that a cow belonging to two brothers, named Kinu and Kālu Ghosh, daily came to Gorāchānd and gave him milk, and that his life would have been saved if he could have sucked unobserved for six days successively. It happened, however, that the milkmen, having failed to get any milk from the cow for four days, watched her, and discovered her in the act of giving suck to Gorāchānd. Thereupon the latter, feeling his end approaching, requested the milkmen to inter his body after his death. He soon afterwards expired, and was buried at Hāroa

The burial of Gorāchānd by the milkmen was observed by another man, who taunted them with the act, and threatened to expose them, so as to put them out of caste. One day the brothers, being unable to bear these taunts any longer, killed the man in a passion, and were taken for trial before Alā-ud-dīn, the Governor of Gaur. The wives of the milkmen went to Gorāchānd's grave and related their misfortunes, when the holy man suddenly rose from the tomb. He immediately repaired to Gaur, arrived before the governor in time to have the brothers released, and returned home with them. Gorāchānd had not forgotten Chandraketu, and in order to bring him into trouble, proceeded a second time to Gaur, and got one Pīr Shāh appointed as Governor of Bālinda. The new governor, soon after his arrival, sent for Chandraketu. The latter obeyed the summons; but having considerable misgivings as to the result, he took the precaution of taking a pair of carrier-pigeons with him, and told his family that, in the event of fortune turning against him, he would let the pigeons fly, and their reappearance at home would be the signal for the female members of his family to destroy themselves. Pīr Shāh harassed Chandraketu so much, that he lost heart and let loose the birds. As soon as his family perceived the return of the pigeons, they drowned themselves. Chandraketu was ultimately released, but on his return he followed the example of his relatives and committed suicide. For a long time the descendants of the brothers Kinu and Kālu Ghosh enjoyed the proceeds of the fair, but the family eventually became extinct, and the tomb is now in the charge of Muhammadans. The Governor Alā-ud-dīn is said to have allotted an estate of five hundred acres of land for the maintenance of the tomb, and certain lands are held nominally for this purpose to the present day.

Hasanābād.—Village in the extreme west of the Basīrhāt subdivision, situated on the west bank of the Ichāmati river. It is the headquarters of a thana and a considerable centre of trade on the Sundarbans boat route. It was for some time the residence of Dr. Carey, the great Baptist missionary. He and John Thomas, who had been a ship's surgeon, landed in Calcutta in November 1793, but after being a month there were reduced to such straits that they had to seek a cheaper locality. Bāndel was fixed upon, but it was ill-suited for Carey's plan of missionary labour, for it afforded him no opportunity of accommodating his habits of life to those of the Indian community, which he considered the most effectual mode of obtaining access to the people. They, therefore, left the place and returned to Calcutta, where they were lent a house in Māniktala by a Bengali money-lender to whom Thomas was in debt. Great was the relief, for Carey's wife had become insane, and her sister and two of his four children were down with desentery. Carey's *munshi* now suggested that the destitute family should move to the waste jungles of the Sundarbans, and there cultivate a grant of land. "With a sum of £16 borrowed from a native at 12 per cent. by Mr. Thomas, a boat was hired, and on the fourth day, when only one more meal remained, the miserable family and their stout-hearted father saw an English-built house. As they walked up to it, the owner met them, and with Anglo-Indian hospitality invited them all to become his guests. He proved to be Mr. Charles Short, in charge of the company's salt manufacture there. Here, at the place named Hasanābād, Carey took a few acres on the Jamuna arm of the united Ganges and Brahmaputra, and built him a bamboo house, 40 miles east of Calcutta. Knowing that the sāhib's gun would keep off the tigers, natives squatted around to the number of three or four thousand."* After a few months Carey left the place for Mālda, where he arrived in June 1794.

Henckellganj.—Village in the Basīrhāt subdivision, situated on the right or west bank of the Kālindi river opposite to Basantpur in the Khulna district. It is one of the chief markets for the *ābāds*, or cultivated clearings, in the Sundarbans, where the inhabitants bring their produce, such as rice, wood and fish, for sale and lay in a stock of tobacco, salt, kerosine oil, etc. The boat route through the Sundarbans leaves the Kālindi here to pass eastwards through the Khulna district. The place is called after Mr. Henckell, Magistrate of Jessore, who was appointed "Superintendent for cultivating the Sundarbans" in 1784. In pursuance of his scheme of colonization, he established three markets for the development of the Sundarbans. Two of these were Kāchua and Chāndkhāli in the Khulna district, and the third was Henckellganj. When Henckell's overseer was clearing the place, which was under jungle when first occupied, the work of reclamation was interrupted by tigers, which made constant attacks on the workmen. The overseer, therefore, called the place Henckellganj in the belief that the tigers would be overawed by the name and cease to molest his men. The name adhered to the village until the survey authorities, in mapping out the district, took the native pronunciation and entered it in the maps as Hingulgunge, so blotting out its history.

Ichāpur.—A village on the bank of the Hooghly in the Barrackpore subdivision, which forms part of the North Barrackpore

* *The Life of William Carey*, by G. Smith, C.I.E., LL.D.

municipality. It is the site of a large Government rifle factory, which employed a daily average of 2,050 hands in 1911. The rifle factory was erected on the site of an old gunpowder factory; on the main gate there is an inscription on a marble slab recording the names of past Superintendents of the gunpowder factory, beginning with Assistant Surgeon Farquhar. The original owners of the site were the Dutch, some of whose buildings still remain. The present buildings were begun in 1903, and electric power was installed in 1905. The first rifle was manufactured, to test the machinery, in 1906, and the regular outturn of rifles began next year. Electrical power is supplied from a large power-house, with seven sets of dynamos and boilers, which also supplies the gun and shell factory: close by are the gas works. The following account of the processes of manufacture is taken from an article called "The Enfield of India," which appeared in the *Statesman* of 27th September 1908.

"A lifetime's use of a rifle, and even an intimate knowledge of its 132 component parts, can give no idea of the immense labour expended in its manufacture, and the hundreds of operations and tests through which it passes, before it is placed in the hands of the soldier or volunteer for service. Every part, down to the smallest screw, passes through many hands and over many machines between the stages of raw material and completion, and each part is disposed of with scrupulous care. To take the case of the rifle body—the steel piece between the fore end and the butt—this component, between the smithy and the final inspection, passes through no less than 208 distinct operations on as many machines, and at the various stages is tested on 102 separate occasions. It is seen first in the smithy as a shapeless glowing lump of steel, and then from hand to hand and machine to machine it passes until it appears in its final shape, but without the polish and browning. These come later, and are of course of no importance whatever in comparison with the work of seeing that it is made on standard lines, and capable of being instantly and perfectly fitted to the other component parts. The bolt, breech, trigger-guard, bolt-head, etc., all pass through many operations and severe tests, before being finally taken to the 'Assembly' section where the rifle is put together. From thence it is forwarded to the 'Inspection'; and, if it passes the practised and vigilant eye of the non-commissioned officer in charge of that important department, to the range for the final test. Perhaps one of the most interesting operations or, more strictly, long series of operations, is the making of the barrel.

The steel rod is bored and then milled, and after a number of other minor operations, the necessity of which can only be adequately gauged by the expert, the rifling is undertaken. That is in itself a work of much delicacy and occupies some time in its completion. Preliminary tests follow, and if the inner surface is what it should be, it is put into the finisher's hand after which it is handed over to the Sergeant in charge of the 'Inspection.' It is possible that at this stage defects that passed at the preliminary inspection are discovered, and the part is promptly rejected as waste.

"In the wood-work section there is also much of interest to be observed. One sees the butt, hewed by a machine with curious looking teeth out of a block of walnut in something under three minutes: and at another machine the fore end is also cut into shape with similar expedition. Smoothing and polishing are effected with marvellous rapidity, and before it is possible quite to realize that the work has begun, there are the two wooden portions of the rifle ready for the 'Assembly,' but in the meantime they have passed through half a hundred hands. The work has to be done to an absolutely correct point, no allowance being permitted for even a thousandth part of an inch of difference.

"But after all the assembling has been completed, and the gauging and testing carried out, there remains the supreme trial on the range, and here the rifle is subjected to a test, from which it emerges a perfect service weapon, or is sent back to the factory for rectification. The men on the range are picked shots, winners at the principal rifle meeting in India, and the distances at which rifles are tested extend from 100 feet to 600 yards. An ingenious machine is provided at each distance, and on this the rifle is placed: telescopic sights are used, so that the error that might be present, even when a crack shot is firing, is eliminated. The ranges are all under cover; therefore no allowance is necessary for force or direction of the wind, or climatic conditions. In short, the rifle is placed in position under perfect conditions that can never be secured in the open on ordinary occasions. The target if shown to a volunteer or regular shot would be his despair, for it is only two or three inches square, and all the shots fired must strike inside the marked space, or the weapon is put aside as defective. If adjustments are possible, well and good: they are made, and the test begins once more from the beginning; if not, the rifle is finally rejected. But it may be taken for granted that once a rifle has passed this supreme test and is issued for service, it is a perfect article: not less so than if it were made in an old established English or Continental factory.

Indeed, it is claimed at Ichāpur that the gauge or test standard there is far higher than that to be found in some European factories. In many of its part the limits of difference between the absolutely accurate and the actual are nil ; in others a thousandth part of an inch.

" Much also has been accomplished outside the factory proper The Park was extremely unhealthy in the first year or two through foundation digging and the absence of proper drainage. The latter, together with an excellent filtered water-supply, was provided in 1905; the many small tanks have mostly been drained, and kerosine oil is used regularly on the others. Septic tank latrines are used in the factories, and lines for workmen have been provided between the Factories Park and the railway lines. This used to be the unhealthiest part of Ichāpur: cholera was practically endemic; but since the land has been cleared and drained, and a filtered water-supply introduced, cholera has disappeared."

Jaynagar.—Town in the south of the Sadar subdivision, situated on the Adi Ganga, an old channel of the Ganges, 31 miles south of Calcutta. The Kulpi road runs through the town, and it is 6½ miles, by water, from the Magrā Hāt station on the Eastern Bengal State railway. Its population in 1911 was 9,245. It is the headquarters of a thana and a station of the London Missionary Society. It contains an out-patient dispensary (opened in 1899), a high school, a sub-registry office and a Bench of Honorary Magistrates. It has been constituted a municipality, the area in municipal limits being 2 square miles. There are four wards, viz., North and South Jaynagar, and North and South Mozilpur. The municipal income is raised by means of a tax on persons; latrine fees are also levied at the rate of 1½ per cent. on the annual value of holdings. Three *melās* are held, viz., (1) the Doljātra in March, which lasts ten days, (2) the Goshtāstamī in November, which lasts one day, and (3) the Goshtajātra in the middle of April, which also lasts one day. About 4 miles south-west of Jaynagar is Mathurāpur, the headquarters of the thana of the same name.

Kāmārhāti.—Town in the Barrackpore subdivision, situated on the Hooghly river. It is bounded on the north by the South Barrackpore municipality, on the west by the Hooghly and on the south by Barnagore, while the eastern boundary lies a little to the east of the main line of the Eastern Bengal State Railway. Its population in 1911 was 18,015. It was formerly part of the Barnagore municipality, but was formed into a separate municipality in 1899. The area within municipal limits is about 3½ square miles, and there are two wards, viz., (1) Ariādaha and Kāmārhāti and (2) Belgharia and Bāsudebpur. The municipal income is obtained from a rate on holdings assessed at 6¼ per cent. on their annual value ; latrine fees are also levied. The town contains a high school and a large hospital, the Sagār Dutt Hospital, which treats both in-patients and out-patients. There are three factories, viz., the Kāmārhāti Jute Mill, of which mill No. 1 employed a daily average of 3,682 hands and mill No. 2 3,351 hands, in 1911, and the Venesta Factory, for rolling tin, etc., which employed 147 hands.

Within this municipality is the greater part of the village of Dakhineswar, where there is a group of temples called Rāni Rāsmani's Navaratna, after the founder, Rāsmani Dāsi of Jānbazar in Calcutta. These consist of two beautiful central temples, dedicated to Kālī and Krishna, faced by 12 minor shrines in honour of Siva. There is a popular burning *ghāt* at Ariādaha, to which Hindu corpses are brought from long distances.

Kānchrāpāra.—Village in the Barrackpore subdivision, situated 28 miles north of Calcutta. This village, which forms part of the Hālishahar municipality, contains a station on the Eastern Bengal State Railway, and the workshops of the Locomotive and Carriage Department of the railway, which employed a daily average of 2,158 hands in 1911.

Kānchrāpāra is also called Bijpur. Here there was a shrine of Kālī, called Dākāiti Kālī, *i.e.*, the Kālī of dacoits, at which, it is said, dacoits used to offer human sacrifices, to propitiate the goddess, before starting on their raids. Part of the tree under which her image stood still remains : it is worshipped by the people and besmeared with vermilion by barren women, who visit it in the hope of obtaining offspring.

Kānkināra.—Village in the Barrackpore subdivision, situated 22 miles north of Calcutta. It forms part of the Bhātpāra municipality and contains a paper mill and jute mills. *See* the article on Bhātpāra.

Kānthālpāra.—A village in the Barrackpore subdivision, situated about a mile from the Naihāti railway station. It is said to be a noted place of Sanskrit learning and was the birthplace of the great Bengali novelist Bankim Chandra Chatterji. A fair is held here during the Rāsjātra of Madan Mohan, which was established half a century ago by the Mahārājā of Nadia, Sris Chandra Ray.

Khardah.—Village in the Barrackpore subdivision, situated on the bank of the Hooghly, 12 miles north of Calcutta. It forms part of the South Barrackpore municipality and contains the municipal offices and municipal dispensary. There is a station here on the Eastern Bengal State Railway, and bricks and brushes are manufactured on an extensive scale. The place is of interest from having been for some time the home of Nityānanda, one of the greatest of Chaitanya's disciples. To him is ascribed the foundation of the village, the legend being as follows :—

Nityānanda came here to live the life of an ascetic on the bank of the Hooghly. One day he heard the lamentations of a woman, and went to her, when she told him that her only daughter had just died. Upon looking at the body, Nityānanda said that the girl was only sleeping. The mother thereupon made a vow that if he would restore her daughter, he should have her for his wife. The saint immediately revived the girl, and wedded her. Being now a married man, he required a house to live in, and asked the landlord of the place for a plot of land for a site. The latter, to mock him, took a piece of straw (*khar*) and threw it into an eddy (*daha*) of the river, telling him to take up his residence there. Nityānanda's sanctity was such that the eddy immediately dried up, and left a convenient site for a dwelling. Hence the village took the name of Khardah.

From Nityānanda's son Bīrbhadra are descended the Goswāmis or Gosāins of Khardah, who are regarded as *gurus*, or spiritual guides, by the Vaishnavas. Khardah has become a great place of pilgrimage for the sect, and large numbers flock thither on the occasion of the fairs held at the Dol and Rās festivals. There is a fine temple containing an image of Shyāmsundar or Krishna, which is the subject of the following legend.

Three centuries ago a Hindu devotee, named Rudra, who lived at Ballabhpur, near Serampore, had a vision, in which the god Rādhāballabh appeared to him and ordered him to go to Gaur and there obtain a stone which was above the doorway of the palace, which he was to make into an image of the god. Rudra went to Gaur and announced to the minister of the Muhammadan Governor the divine orders he had received. Soon after his arrival the stone began sweating, and the minister, who was a pious Hindu, pointed out to his master that the drops oozing out of it were tears and that so inauspicious a stone should be removed. This was done, but the stone was so heavy that it was difficult to get it into a boat. It fell into the water, and then miraculously floated to Ballabhpur. Rudra made three images out of it, called Rādhāballabh, Shyāmsundar and Nanda Dulāl. Bīrbhadra longed to obtain one of them, but Rudra could not be induced to part with any. One day, however, when Rudra was observing his father's *srāddha*, the sky became overcast, and it seemed certain that the ceremony would be interrupted by rain. Bīrbhadra, who was one of the guests, thereupon began praying, and, though there was heavy rain all round, not a drop fell on the spot where Rudra and his guests were. Bīrbhadra then begged Rudra to make him a present of one of the images, and the grateful Rudra made over to him the image of Shyāmsundar, which he installed at Khardah. The image of Rādhāballabh is at Ballabhpur, and that of Nanda Dulāl at Shāhibāna, a village 5 miles east of Barrackpore. It is considered an act of great religious merit to visit all three images in one and the same day. A short distance from the Vaishnava temple at Khardah there is a cluster of 24 shrines dedicated to Siva.

Kulpi.—Village in the Diamond Harbour subdivision, situated a mile inland from the bank of the Hooghly, about 33 miles (by water) south of Calcutta. It is the headquarters of a thana, and is connected with Vishnupur (Bishtupur) by a road 11½ miles long. The reach of the Hooghly here is a favourite anchorage for vessels proceeding up and down the river. It is described as follows in Hamilton's *East India Gazetteer* of 1815:—

"The shores here are a bed of mud, and the banks of the river covered with trees and thick jungle. Opposite to the anchorage of the ships, which lie about half a mile from the shore, is a creek, and at a mile from its entrance stands the town of Culpee. The crews of the ships stationed here suffer dreadfully from its extreme unhealthiness, numbers daily falling sacrifices to the pestilential exhalations from the rotten jungle and mud."

Magrā Hāt.—Village in the Diamond Harbour subdivision, situated 25 miles (by rail) south of Calcutta. It is the headquarters of a thana and contains a dispensary and sub-registry office. It is also a station of the Society for the Propagation of the Gospel, which has a church here capable of holding 150 persons. Owing to its position upon the confluence of important waterways and upon the Diamond Harbour branch of the Eastern Bengal State Railway, Magrā Hāt is the largest rice mart in the south of the district, and agencies have been opened for the purchase of rice and the sale of kerosine oil.

Māniktala.—Town in the Sadar subdivision situated to the north-east of Calcutta, from which it is separated by the Circular and Baliāghāta Canals. Its population in 1911 was 53,767, of whom 32,921 were Hindus and 20,656 were Musalmāns. The increase since 1901 amounted to 21,380, or 66 per cent., which is accounted for by immigration, for the number of immigrants, *i.e.*, persons born outside the 24-Parganas and enumerated in the town, rose by 19,296. Apart from the attraction of labour to industrial concerns, there have been two subsidiary causes of this increased influx. In the first place, the dismantling of *bastis* in the fringe area of Calcutta, the opening out of new roads and the construction of new sewer lines caused a shifting of population, the people who had been unhoused crossing the canal into Māniktala. In the second place, the comparatively low price of land and the convenient situation of the town induced a certain number of the residents of Calcutta to make their homes in Māniktala, including some who desired to forestall the rise in the price of land which was anticipated from the Calcutta Improvement scheme. The immigrants now represent three-quarters of the population, and as they are nearly all males, there is a striking disproportion of the sexes, there being only 7 females to every 10 males. The density of population is 25 persons per acre, the average being 20 in Ward I, 24 in Ward II, and 26 in Ward III.

The town is an industrial suburb of Calcutta, wedged in between the Circular Canal on the west, the New Cut Canal on the east, and the Baliāghāta Canal on the south. Baliāghāta in the south of the town is the seat of an extensive trade in rice imported from the eastern districts of Bengal, while along the frontage of Circular Canal a brisk business is done in firewood, loose jute and rice. Ultadānga and Nārikeldānga are two other quarters in which there are a number of manufacturing works, of which the largest is the Soora (Sura) Jute Mill; this employed a daily average of 1,176 hands in 1911. The nursery gardens of two Calcutta florists are also situated in the town.

Māniktala was included in the Suburban municipality until 1889, when it was made a separate municipality. The area within municipal limits is 3½ square miles, and there are three wards. The municipal income is obtained chiefly from a rate on holdings assessed at 7½ per cent. on their annual value. There is a lighting-rate assessed at 3 per cent. on the annual value of holdings, and a water-rate assessed at 5 per cent. in the case of those that have no filtered water connection, and at 7 per cent. in the case of those having connection. Latrine fees are also levied.

Just beyond the western boundary is a Jain temple, which is described as follows in *Bengal Past and Present* of April 1908:— "Branching off from Upper Circular Road, in an easterly direction, runs a street which takes its name from the temple and conducts the visitor to a fine gateway. The shrine itself, in the Jain style of architecture, is dedicated to Sītal Nāthji, the tenth of the Tirthankaras or Jain prophets. A flight of marble steps leads up to the temple, the most beautiful in Calcutta, round three sides of which runs a verandah. The interior of the building is profusely ornamented, the mosaic decoration, as well as the glass and stone work, being specially noticeable. A chandelier with a hundred and eight branches embellishes the sanctuary. There is also an elegantly furnished parlour, as well as reception rooms, guest-houses, and other accommodation. No description of the place would, however, be complete without some mention of the delightful grounds. In addition to the usual features of well-kept gardens, there stands in the centre an artistic fountain, while around are interspersed garden benches and statuary. Apart from testifying to the wealth of the Jain community, which would surely have abundant reason to feel proud of these palaces of glittering sunshine, the building, with its surroundings, is a standing monument to the good taste of the venerable and charitable founder who built it forty-one years ago."

The neighbourhood of Māniktala was the scene of a battle, on 5th February 1757, between Clive and the forces of Sirāj-ud-Daula, of which an account will be found in Chapter II.

Naihāti.—Town in the Barrackpore subdivision, situated 24 miles north of Calcutta on the bank of the Hooghly river. Its population in 1911 was 18,219. It contains a station on the Eastern Bengal State Railway, and is the junction of a branch line across the Hooghly, which connects with the East Indian Railway. The bridge across the Hooghly is called the Jubilee bridge, because it was opened in 1887, the year of Queen Victoria's Jubilee. It is a fine structure built on the cantilever principle with a length, between abutments, of 1,200 feet divided between three spans. The central span, which is 360 feet long, rests on two piers in the middle of the river: the other two (each 420 feet long) project from either bank. The piers are sunk to a depth of 73 feet below the bed of the river on iron caissons. The height of the bridge above the highest water mark is 36½ feet, so that there is ample space for the passage of river steamers and native cargo boats. All heavy goods traffic from the west of the Hooghly intended for export from Calcutta, such as coal, wheat, etc., passes over this bridge to Naihāti and thence, by the Kānkurgāchi loop, to the Kidderpore docks.

Naihāti was constituted a municipality in 1869, but the area within its limits has been much curtailed by the separation of the

Bhātpāra municipality in 1899 and of the Hālishahar municipality in 1903. The municipal income is obtained chiefly by means of a tax on persons, which is assessed at the rate of Re. 1-2 per Rs. 100 of income. Latrine fees are also levied at the rate of 5 per cent. on the annual value of holdings.

There are two large jute mills in the town, viz., the Naihāti Jute Mill, which had a daily average of 2,336 employés in 1911, and the Gauripur Jute Mill, in which the average number was no less than 8,445. The latter mill is so called after the village of Gauripur or Garīfa, which is within municipal limits; this village was the birth-place of Keshab Chandra Sen, the founder of the Brâhmo Samāj (born in 1838). Prior to the creation of the Bhātpāra municipality, there were six emigration depôts in the town, it being a halting place for coolies proceeding to Assam; but five of these were in Bhātpāra, and only one was left in Naihāti. The town is the headquarters of a thana and has a Bench of Honorary Magistrates, a sub-registry office, a high school and a dispensary, which is maintained by the municipality.

Palta.—Village in the Barrackpore municipality, which forms part of the North Barrackpore municipality. It contains the water-works of Calcutta, of which the following account is taken from an article that recently appeared in the *Statesman*:—

"In England, the supply of water for large towns and cities is usually collected in hills away from the town, and sometimes has to travel very long distances before reaching the consumer—the Birmingham catch-ground, for instance, is away in the hill of Wales. This system was, however, impossible in Calcutta, for there is not a hill one hundred feet high within one hundred miles of the city. Another source of supply had, therefore, to be found, and it was decided by the authorities about half a century ago to take the water from the river Hooghly. There is nothing of the crystal spring of which the poet sings about the Hooghly, and in the city itself the water is so brackish that it was felt that it would be impossible to purify it. Therefore, a place had to be found where the water contained a minimum of silt and other impurities, and Palta was eventually fixed upon. Here the water is pumped into settling tanks and from there gravitates on to filter beds, whence it passes into the pipes and is pumped into the city.

"It might be thought that the system required to cleanse and purify water so dirty as that which comes from the Hooghly would be a very elaborate one, but it is not. On the contrary it is exceedingly simple. From the river, as has been said, the water is pumped straight into the settling tanks, where it stays until the silt and other heavy matter settles to the bottom. A day, or a day and a half, is sufficient for this process, and the surprising thing is that nature itself does the work, except in the rains. In this period the river water is simply laden with silt, and allumino-ferric is put into the settling-tanks to help it to settle, but throughout the rest of the year nothing is put in, and the foreign matter in the water settles naturally. Some idea of the quantity of silt which the water contains can be gained from the fact that last year one tank was cleaned, and over four hundred thousand cubic feet of silt was taken out. The tank had been in use then for two years. The partly-purified water then passes by gravitation on to the filter beds, of which there are 42 constantly in work. These filters cover an area of 850 square feet, and have to filter the whole of the water-supply for Calcutta. They are exceedingly simple in construction, and it is little short of marvellous that they do the work so well, and so effectively remove all the impurities from the water. The filter is composed of a layer of pebbles, with 4 inches of coarse Magra sand on top, and 2 feet 6 inches of river sand above that. The water goes on to the filter beds still yellow and dirty, although the silt has been removed; it comes out, after having filtered through the sand and the pebbles, absolutely clear and pure. It goes through the filters at the rate of 4 inches per hour. The simplicity of the process, considering the work that has to be done and the uniformly successful result, is surprising, but constant care is necessary to see that the filters do their work properly. The sand, which is the all-important factor in the filter—for it is it that does all the cleansing of the water—has in its turn to be cleansed by a supply of the very water which it has purified. The method of washing the sand is very primitive. It is thrown into a square brick chamber, with a false bottom pierced with very small holes. Water is introduced under pressure, and coolies, standing up to their middles, agitate the sand with shovels until it is clean. Those in charge of the works have long objected to this system, and have introduced a more up-to-date method on a small scale, which will doubtless soon replace the older method altogether. In the new method the sand is revolved by machinery. The water from each filter bed is tested once a week. When the water comes from the filter-beds it is collected in one central well, and then pumped into the two big mains which run, one on each side of the Barrackpore road, to Tāllah, where it is stored in underground reservoirs, which have a capacity of eight million gallons, until required in the town mains.

"The pumping machinery at Palta is very powerful. The engine which pumps the water from the river into the settling tanks is capable of lifting two million gallons an hour, and it does more work than three engines used to do; while the engine which pumps the water from the filter-beds to Palta deals with the whole supply to the city—which is now between 35 and 37 million gallons daily.

"An article on the water-supply would not be complete without a reference to the overhead tank at Tāllah. This tank was designed by Mr. W. B. MacCabe to act as a balancer. The pumping engine at Palta is not capable of sending down sufficient water to meet the demand in the middle of the day, when the consumption is very heavy, and it is then that the overhead reservoir comes into operation. When sufficient water is not coming down to fill the town mains, water from the tank automatically flows into the mains and so keeps the supply equal to the demand. The tank is refilled again during the night hours when less water is being used."

Pānihāti.—Town in the Barrackpore subdivision, situated on the bank of the Hooghly, 9 miles north of Calcutta. Its population in 1911 was 11,118. It was formerly part of South Barrackpore, but was constituted a separate municipality in 1900. The municipal income is derived from a tax on persons assessed at 12 annas per hundred rupees of income; latrine fees are also levied at the rate of 7 per cent. on the annual value of holdings. There are two annual fairs, viz., the Rāsjātra *melā* in November, which lasts four or five days, and the Baishtabi *melā* in May, which lasts one day.

The village of Agarpāra, which lies within municipal limits, contains a church capable of holding 500 people, with a tower 74 feet high, which was built in 1837 by Mrs. Wilson; there are also a female orphanage and school under the management of the Church Missionary Society. A fair, called the Tārāpukur *melā*, is held here at the end of January, and lasts one day.

Near the railway station at Sodepur (1 mile north of the Agarpāra railway station) there is a Pinjrāpol, or home for aged and diseased animals, which is maintained by subscriptions, chiefly from the Mārwāri community of Calcutta. Here a fair, called the Pinjrāpol Gopāstomi *melā*, is held in November every year; it was started about 1890, and lasts one day. There were formerly glass works at Sodepur, which were worked by the Bengal Glass Co. from 1891 to 1902, when the undertaking was abandoned. There is a high school in Sodepur and another in Pānihāti.

Rājpur.—Town in the Sadar subdivision, situated on the road from Calcutta to Kulpi, 11 miles south of Calcutta. Its population in 1911 was 11,607. The town constitutes a municipality with an area of 2 square miles and five wards, viz., (1) Rājpur, (2) Harināvi, (3) Kodālia and Chāngripota, (4) Mālancha and Māhinagar, and (5) Elāchi and Jagadal. The municipal income is derived from a tax on persons assessed at 10 annas to Re. 1 per hundred rupees of income according to the circumstances of the assessees; latrine fees are also levied at the rate of 6¼ per cent. on the annual value of holdings. The water-supply is obtained almost entirely from tanks. The portion of the town north and west of the Kulpi road drains into the Adi Ganga, and that lying east and south of the road into the Arahapanch Khāl. There are a high school and a municipal dispensary at Harināvi. Three *melās*, each lasting a day, are held, viz., the Rāsjātra in April, the Gostabihār in March or April, and the Snānjātra in May or June.

Sāgar Island.—Island in the Diamond Harbour subdivision, situated at the mouth of the Hooghly river. It is bounded by the Hooghly on the west, and by the Bārātala or Channel Creek on the east, while the Bay of Bengal washes its southern face. The northern extremity of the island, which is about 25 miles long, is called Mud Point and is the site of a telegraph station. Here the Sundarbans steamers leave the Hooghly and pass down Channel Creek on their eastward route through the Sundarbans. At the south-western angle of the island stands the Sāgar lighthouse, which was built in 1808, and the southern sea face is the site of the great bathing festival of Gangā Sāgar.

Situated as it is, at the point where the holy Ganges once mingled its waters with the sea, it is regarded as a peculiarly sacred spot. The legend accounting for its sanctity is as follows:—

Sagar, King of Oudh, the thirteenth ancestor of Rāma, had performed the *Aswamedha jajna*, or horse-sacrifice, ninety-nine times. This ceremony consisted in sending a horse round the Indian world, with a defiance to all the earth to arrest its progress. If the horse returned unopposed, it was understood to be an acquiescence in the supremacy of the challenger, and the animal was then solemnly sacrificed to the gods. When King Sagar made preparations for the hundredth sacrifice, Indra, King of Heaven, who had himself performed the ceremony a hundred times, jealous of being displaced by this new rival, stole the horse, and concealed it in a subterranean cell, where the sage Kapila, or Kapilmuni, was absorbed in meditation, dead to all occurrences of the external world. The sixty thousand sons of Sagar traced the horse to his hiding-place, and, believing the sage to be the author of the theft, assaulted him. The holy man being thus roused, opened his eyes and cursed his assailants, who were immediately burnt to ashes and sentenced to hell. A grandson of Sagar, in search of his father and uncles, at last came to Kapilmuni, and begged him to redeem the souls of the dead. The holy man replied that this could only be effected if the waters of Ganga could be brought to the spot to touch the ashes.

Now Ganga was residing in Heaven, in the custody of Brahma the Creator, and the grandson of Sagar prayed him to send the goddess to the earth. He died, however, without his supplication having been granted. He left no issue; but a son, Bhagīrath, was miraculously born of his widow, and through his prayers Brahma allowed Ganga to visit the earth. Bhagīrath let the way as far as Hāthiāgarh, in the 24-Parganas, near the sea, and then declared that he could not show the rest of the way. Whereupon Ganga, in order to make sure of reaching the spot, divided herself into a hundred mouths, thus forming the delta of the Ganges. One of these mouths reached the cell, and, by washing the ashes, completed the atonement for the offence of the sons of King Sagar, whose souls were thereupon admitted into heaven. Ganga thus became the sacred stream of the hundred mouths. The people say that the sea took its name of Sāgar from this legend; and the point of junction of the river and sea at Sāgar Island is a celebrated seat of Hindu pilgrimage, to which thousands of devout pilgrims repair every year during the great bathing festival.

The festival is held at the time of the Makara Sankrānti, when the sun enters Capricorn, which is identified with the 1st Māgh (in the middle of January), and is the occasion of a large fair. The fair takes place on the bank of a small creek leading to the sea, on a piece of sandy ground, where mat booths are run up for the sale of the hawkers' wares. The fair lasts several days, but three days are the limit of the religious festival, The first ceremony is the propitiation of the ocean, by casting into it various offerings with short ejaculatory prayers; the oblations are commonly cocoanuts, fruits, or flowers. The most appropriate gift is that of the five gems (*pānch ratna*), consisting of a pearl or diamond, an emerald, a topaz, and a piece of coral, along with a cocoanut, an areca-nut, and the sacred thread worn by Brāhmans. These are wrapped up in a cloth, and cast into the creek which communicates with the sea, and also at the confluence. The jewels are in general not worth more than a rupee or two. On the first day the pilgrims bathe in the sea early in the morning, some repeating the bath at noon. Some also have their heads shaved after bathing, and many of those whose parents are recently deceased celebrate the *srāddha*, or obsequial ceremonies, on the sea-shore. After ablutions, the pilgrims repair to the temple dedicated to Kapilmuni. On the second and third days, bathing in the sea, adoration of Ganga, and the worship of Kapilmuni continue as on the first, after which the assemblage breaks up. During the whole time the pilgrims, for the most part, sleep on the sand, for the crowds are so great that they could not all find accommodation in the boats.

The image of Kapila is a shapeless block of stone daubed with red paint. During the greater part of the year it is kept in Calcutta; but a week or two before the festival it is handed over to the priests, who take charge of it during the festival, and receive a share of the pilgrim's gifts. It is placed in a temporary temple, as the old one has been washed away by the encroaching sea, and stands on a platform of sand about four feet high. A bamboo railing in front keeps off the crowds, who go past it from day light till dark. According to Wilson—"In front of the temple was a banian (*bar*) tree, beneath which were images of Rāma and Hanumān. The pilgrims commonly wrote their names on the walls of the temple, with a short prayer to Kapila, or suspended a piece of earth or brick to a bough of the tree, with some solicitation, as for health, or affluence, or offspring, and promised, if their prayers are granted, to make a gift to some divinity. Behind the temple was a small excavation termed Sitākund, filled with fresh water, of which the pilgrim was allowed to sip a small quantity, on paying a fee to the manager of the temple. This reservoir was probably filled from the tank, and kept full by the contrivances of the mendicants, who persuaded the people that it was a perpetual miracle, being constantly full for the use of the temple."*

The estimates of the number of persons attending the festival vary very widely. Fifty years ago Wilson wrote:—"Some years ago they were considered to average about 100,000; but I have been informed by high authority that latterly the number has increased to double that amount." Hunter put the number somewhat vaguely at "hundreds of thousands," and Wilkins in *Modern Hinduism* at the more modest figure of 150,000. On the other hand, Mr. F. E. Pargiter, I.C.S., then Commissioner in the

* H. H. Wilson, *Essays on the Religion of the Hindus* (1862, Vol. II).

Sundarbans, in an article on the Sundarbans published in the *Calcutta Review* for October 1889, wrote:—"The festival is decaying, unless excursion steamers should resuscitate it as a pleasure trip; and the numbers who attend it are far below the estimates often made. I doubt if the number exceeds 5,000, though it is popularly stated to be something like ten times as many." The festival appears to have gained greater popularity since Mr. Pargiter wrote, and the number attending it is now estimated at 30,000 to 50,000.

Formerly suicides and the destruction of children were features of the festival, many of the pilgrims making voluntary sacrifices of themselves or throwing their children to the sharks and alligators. "On shore," it was said, "the jungles swarm with tigers of the largest and most ferocious sort, so that both elements are equally dangerous."* It is said that, in 1801, 23 persons were exposed or drowned in one month, but next year this horrible practice was suppressed by the Marquess Wellesley. It was not, like the oblation of fruits or jewels, intended to obtain the favour of the deified ocean, but in satisfaction of a vow. For instance, a childless woman would make a vow to offer her first born at Gangā Sāgar, in the hope that such an offering would secure for her additional progeny.

The reclamation of the island from jungle was started early in the nineteenth century. In 1811, a Mr. Beaumont applied for permission to hold a hundred acres of land in the island for the purpose of establishing a manufactory of buff leather, and asked that all tiger-skins brought to the Collector's office might be made over to him for this purpose. His application for land was granted by the Board of Revenue in November 1811; and in the following year, in consequence of a Government resolution offering favourable terms for the cultivation of Sāgar Island, Mr. Beaumont applied for a grant of land on a cultivating tenure. This application was rejected on the ground that Government had decided not to grant leases to Europeans for cultivation. Leases of the island were offered to Indians only, and many proposals were received from them, but this scheme of colonization was a complete failure. The island was subsequently leased to an association composed of Europeans as well as Indians, free of rent, for thirty years, and to pay only four annas per bigha ever after. The undertaking was begun with vigour, but so many unforeseen difficulties occurred that up to the 1st September 1820 not more than four square miles had been

* Hamilton's East India Gazetteer, 1815.

effectually cleared. Amongst other obstacles it was found that as the woods were cut down, the sea encroached, the sandy beach not having sufficient tenacity to resist its invasion. Twenty-five families of Maghs from Arakan were settled at the confluence of two creeks, and a road constructed for the accommodation of pilgrims to the temple of Kapila.*

In 1819, Mr. Trower, Collector of the 24-Parganas, originated a company, called the Saugor Island Society, for the systematic reclamation and development of the island; he himself was a considerable shareholder, and the central part of the island was called Trowerland after him. The company obtained a grant of the whole island, subject to certain conditions (the breach of which entailed forfeiture of the grant) and carried on operations vigorously until 1833, when their work was destroyed by a cyclone and they abandoned the project. Their interest in the northern part of the island was then taken over by four European gentlemen, who combined the manufacture of salt with the cultivation of rice. The progress of the island was again interrupted by the cyclone of 1864, when 4,137 persons or three-fourths of the population perished, only 1,488 being left. Since then considerable progress has been made in reclaiming the waste, and the north of the island is under cultivation, but the south is still dense jungle.

Salt Water Lake.—Swamp in the Sadar subdivision, situated about five miles east of Calcutta, with an area of about 30 square miles. This is a low depression, which is being gradually filled by the silt deposits of the tidal channels that intersect it; a portion, at Dhāpa, is also being reclaimed by the deposit of the street refuse of Calcutta, which is conveyed there daily by a municipal railway. The lake formerly extended much further west, and in 1757 came within a mile of the Marātha Ditch (Lower Circular Road).

Shāmnagar.—Village in the Barrackpore subdivision, situated on the bank of the Hooghly, 19 miles north of Calcutta. It forms part of the Gārulia municipality and contains a station on the Eastern Bengal State Railway. A short distance east of the station are the ruins of a mud fort, surrounded by a moat, four miles in circumference, which is said to have been built in the eighteenth century by the then Rājā of Burdwān as a refuge from the Marāthas. It now belongs to the Tagore family of Calcutta, and its ramparts are studded with thick date plantations. A Sanskrit college and a charitable dispensary are maintained by

* Hamilton's East India Gazetteer, 1828.

the Tagore estate. The village contains several large industrial works, viz., the Shāmnagar Jute Mills, of which one employed 4,547 hands and the other 1,759 hands in 1911, the South Alliance Jute Mill with 2,040 hands, the Dunbar Cotton Mill with 910 hands, and the Bhajoram Jute Press, which did not work in 1911.

South Suburbs.—Municipality in the Sadar subdivision, which, as the name indicates, lies in the suburban area of Calcutta. It extends round the "Added Area" from near Sealdah on the north-east to the Tollygunge municipality on the south. Its population in 1911 was 31,533. The area of the municipality has been much reduced in recent years by the separation of Garden Reach in 1897 and of Tollygunge in 1901. A great portion is rural in character, for it consists of a number of scattered suburban villages interspersed with paddy fields and other arable land. The villages are usually built on more or less elevated ground, and the drainage finds its way into the fields below them. The municipal income is derived chiefly from a rate on holdings assessed at 7½ per cent. on their annual value; latrine fees are also levied according to a prescribed scale, and there is a water rate assessed at 3 per cent. on the annual value of holdings. The most important of the villages making up the municipality are Barīsa and Behāla. In the former there is a municipal dispensary. In the latter, Mānik Chānd, the Dīwān or Minister of Sirāj-ud-daula, who was appointed Governor of Calcutta after its capture by Sirāj-ud-dāula in 1756, had a country seat surrounded by a large garden. In this garden, which may still be seen on the Diamond Harbour Road, 4½ miles from Calcutta, Mānik Chānd encamped during the siege of the city.

Tāki.—Town in the Basīrhāt subdivision, situated on the Jamuna or Ichāmati river on the eastern boundary of the district. Its population in 1911 was 5,202. It forms a municipality with an area of two square miles divided into four wards, viz., (1) South Tāki, (2) North Tāki, (3) Saiyadpur and (4) Beokāti and Jalālpur. The municipality is practically nothing more than a collection of villages interspersed with rice fields. It contains a Government high school and a charitable dispensary called the Tārā Sankar Chaudhri's dispensary. The municipal income is raised by a tax on persons assessed at 1 per cent. of annual income.

The town is situated on the Bārāset-Basīrhāt Light Railway, 41 miles from Calcutta. It is 72 miles from Alipore *viâ* Tolly's Nullah and Canning (being 40 miles from Canning), 54 miles from Takta Ghāt at Hastings *viâ* the Chitpur and Bhāngar canals, and 7 miles from Basīrhāt by a *kutcha* road.

Titāgarh.—Town in the Barrackpore subdivision, situated on the bank of the Hooghly, 13 miles north of Calcutta, with a station on the Eastern Bengal State Railway. It was constituted a municipality in 1895, being detached from South Barrackpore, which adjoins it on the north, south and east. It had a population of 45,171 persons in 1911. The area within municipal limits is about 1½ square miles, and there are four wards, viz. (1) Dās Bāgān, (2) Mill Bazar, (3) Old Barrackpore and (4) Khardah. The municipal income is obtained from a rate on holdings assessed at 7 per cent. on their annual value; latrine fees are also assessed at 5½ per cent. on the annual value of holdings and at 9 pies a month per head in cooly lines.

At the beginning of the nineteenth century Titāgarh was a favourite place of residence for Europeans, several of whom had country houses in it. It contained a dockyard, from which was launched a vessel of 1,445 tons, the *Countess of Sutherland*, said to have been the largest merchant vessel ever built along the Hooghly. At present Titāgarh is a busy industrial centre with several jute mills and a paper mill as shown in the margin.

Name of mill.		Average number of employés (1911).
Kelvin jute mill	...	2,765
Khardah ditto,	No. 1 ...	4,405
Do. ditto,	No. 2 ...	5,103
Kinnison ditto	...	5,339
Standard ditto	...	3,849
Titāgarh ditto,	No. 1 ...	5,637
Ditto ditto,	No. 2 ...	6,537
Titāgarh paper mill,	No. 1 ...	1,423

Tollygunge.—Municipality in the Sadar subdivision, situated immediately south of and adjoining Calcutta. Its population in 1911 was 18,433. It contains a police-station, the barracks of the 24-Parganas police reserve, the golf links of the Royal Calcutta Golf Club, the grounds of the Tollygunge Club (containing another golf course) and a steeplechase course. Several of the descendants of Tipu Sultān's family have their residence here.

Tollygunge was included in the South Suburban municipality until 1901, when it was constituted a separate municipality. The municipal income is raised by means of a rate on holdings assessed at 7½ per cent. on their annual value. Latrine fees are levied according to a prescribed scale, and there is also a water rate assessed at 5½ per cent. on the annual value of holdings within 400 feet of the nearest hydrant and at 4½ per cent. in the case of holdings situated beyond 400 feet but within 1,000 feet from the nearest hydrant.

The place is so called after Colonel William Tolly, who, as related in Chapter X, canalized an old channel of the Ganges, which was, and is, consequently known as Tolly's Nullah. The first we hear of this officer is that in 1766 Captain Tolly was busy with a factory on the border of the Sundarbans. In 1776 he began excavating Tolly's Nullah, which appears under that name in a list of the boundaries of Calcutta in 1794, and in 1780 he purchased Belvedere from Warren Hastings. The last mention of him occurs in the *Calcutta Gazette* of 23rd September 1784, where it is stated that Colonel Tolly died on the voyage to St. Helena.

INDEX.

A.

B.

K.

L.

M.

N.

O.

P.

R.

S.

T.

U.

V.

W.